Symmetry and Balance book collection, division of time book I and book II
The underlying structure of physics

By

Tahir Iqbal (c) 2020

We take as our paradigm the concept of all is flux from Heraclitus of antiquity. He wrote that no one steps in the same river twice, as you step again in a river it is a different river since it has flowed away. The paradigm of post Platonic thought is causal laws, A causes B. Heraclitus wrote of fire caused by air which while earth causes water, yet water causes fire and fire causes water. This is the concept of interdependent equations. Instead of the single causal variable there is we find a curious generality among complex systems, nonlinear dynamics, chaos theory and random systems. More can be said than existing mathematics.

An argument in the debate is that this is not scientific, so we show the deficiencies of physics in our underlying theory of physics; both theoretical and in numerical estimation. I have an alpha for that. The underlying theory we espouse is what is important in those systems and structures of theories.

We use the concept of Lie groups and algebras in our analysis of solving interrelated equations (themselves expressed as matrices). A point against this is the sparse matrix (the zero in what we call the jacobian of a system of equations. But look at population growth, a zero in this makes no population or growth, or a fixed population not growing but becoming zero.

The ideology of 'free maths' is used in our methodology, that is where maths does not have to follow a set authoritarian prison in the mind. Instead you have your own personal theory you learn and fit into the literature of mathematics. We believe this increases the amount of mathematical knowledge and interest in the universe, and making learning maths a fun and free experience. I owe this concept to my contemporary, Rory Madden.

We use the ancient Greek thought of the labyrinth, where the mathematician goes through multiple routes of deduction and proof of mathematics. We simply argue that no one is wrong and there is no wrong mathematics, so each person can build any number of ideas. The key is connecting different parts of math to each other, and using a Kantian of invention and simulation to separate good theories from less advanced ones. Invention means making new technologies and products from theories, whilst simulation means using computing to model and explore mathematics. Instead of control and experiment as the Kantian (which is only useful for A causes B, rather than A,B,C each causing each other), you have a system of thought that can deal and does not dismiss interdependency and system of equations.

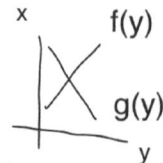

graph 1.01
the underlying structure of physics

GRAPH 1.01, THE UNDERLYING STRUCTURE OF PHYSICS

The underlying structure of physics is given in graph 1.01

1. There are two functions at the base of physics. Typically we see a constraint in equations that model real phenomena. This is expressing a line for an equation of x+y= a, where a is a constant. This is then put alongside another equation, of x multiplied by y for the same variables in the x+y=a line.
2. Given monotonicity of the functions there is an equilibrium. This work is not to show that such a structure is wrong, rather it reveals deeper questions in the nature of reality. The x multiplied by y curves are convex in the sense of having a d2y/dx2 function that is increasing as you go above or below the equilibrium.
3. The reason for this is that say there is an upward sloping function, f(y) and a downward sloping function, g(y), where the equation, x+y=a is manifested. As they are a constant , a (which shows the range of f(y) and g(y) at its upmost as being a, and at its lowest being a on the other side (by means of solving the x+y=a equation so y = x-a) then there is a function of f(y)-g(y) that defines the equilibrium where they cross, in other words a zero at the point of half of a (the constant).
4. Since x multiplied by y grows at a faster rate than the difference, f(y)- g(y) there is an a curve often called an 'indifference curve', which we implore we should not be indifferent of.
5. Examples of the underlying structure in physics:
6. 1. Potential energy and kinetic energy having a constant (in our analysis the constant s) due to conservation of energy, then the multiplication of potential and kinetic energy to give the work done by the system (and is the multiplication of the x and y applied by graph 1.01), as well as force = mass multiplied by acceleration with force constrained by a constant equation, for example of potential and kinetic energy sums.

7. 1. Sinks and sources in fluid mechanics being summed together to be the same (in other words source and sink in addition to make the conservation of the variables, the addition line of x+y=a (x is source and y is sink). In optimal control theory this is expressed as source as initial condition and sink as terminal time. This is the conservation principle in optimal control.

8. 1. Quantum mechanics, where there is a conservation of probability as the x+y=a constraint.

9. 1. The first law of thermodynamics gives a conservation law, which can be expressed as a x+y=a so there is the constraint.

10. 1. Hookes law, where the constraint is the difference between the spring and its original length, then put into a multiplication of this length and the constant to give a force, to give an equilibrium which is governed by the underlying structure of physics.

11. 1. Newtons law of gravity: the multiplication is the mass of the two bodies, while the constraint is the distance between the two centres of the bodies attracting in gravity

12. 2. Newton's third law of action and reaction being equal and opposite is another example of the underlying structure of physics as the two functions are placed into this.

13. 1. Coulombs law has the same structure as 11.1 Newton's law of gravitation. Once again we see the underlying structure of physics.

13. 1. In maxwells equation there is the assumption of conservation of charge. Interestingly this moves through the underlying model of physics with the multiplication existing in many different variables, through Ampere's law and Gauss's law. Putting it in layers of the same underlying structure makes little difference to finding the equilibrium and determining dynamics.

14. 1. Faraday's law of induction has the constraint of x+y = a by using a line integral of an electric field. This is important as we have to generalise our work to more than 2 dimensions to be able to apply this underlying structure throughout modern physics where there are matrices, vector calculus and stochastic phenomena.

15. 1. The use of Hamiltonians in models of physics can be seen to fall into our underlying structure of physics as in canonical formulations there is a connection and mirror symmetry in the Hamiltonian, so there is an equilibrium as there are two functions that are opposite in sign of themselves (so going back to graph 1.01). As a hamiltonian with lagrangian, another kind of solution, you can look at the Hamiltonian to be:

$$\langle x. y \rangle - (x+y = \text{constant}) = 0$$

The inner product of the two functions and their sum and constraint from a constant as well as zero, is the same as our symmetry and balance idea, so fits with the underlying structure of physics (graph 1.01) as $\langle x.y \rangle$ is f(y) and (x+y= constant) is g(y) in this graph. There are really only a few possibilities for equations; zero, 1, limit to infinity, limit to minus infinity, oscillations and special solutions in a certain domain (like Kammaa equations).

Exercise

Using the aforementioned argument, draw a set of indifference curves and constraints for a 3 dimensional Hamiltonian. Here you will see that there is another possibility for the equations, that of multiple solutions and dynamics of choice, depending on which route you go through in the labyrinth of mathematical deduction.

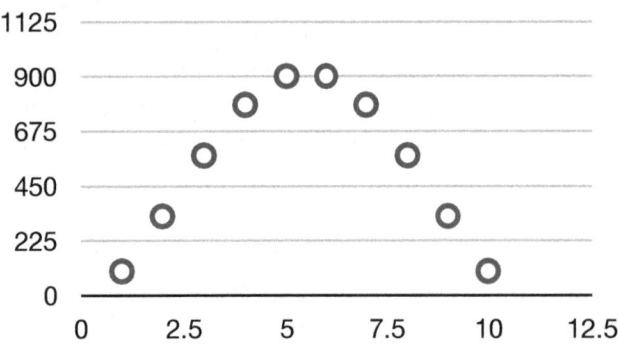

Graph 1.1 the parabola of multiplication

The graph 1.1 shows what the result of equations multiplied by each other (from the underlying structures discussed in the many examples in point 5 above). They have an equilibrium at half the end of the range of the variables. So where a in the constraint, x+y=a, is given, then the optimum is at a divided by 2.

The underlying theory of physics can be seen in the following table, table 1:

Table 1

Potential energy	Kinetic energy	Potential energy x kinetic energy
10	0	0
9	1	9
8	2	16
7	3	21
6	4	24
5	5	25
4	6	24
3	7	21
2	8	16
1	9	9
0	10	0

We call potential energy , PE, and Kinetic energy, KE.

There is the x+y=a constraint, so PE= 10 where KE=0, this goes up for KE while PE goes down, due to the constraint, or conservation principle, used much throughout physics. Then there is the multiplication part, for example work done being proportional to PE x KE. The third column, PE x KE is an expression of the parabola we showed earlier when simply written as the KE as x axis. It is easy to see that the optimum, equilibrium, fixed point, zero and gradient zero of a tangent to the parabola, are at PE= KE= 5, where 10 is the top of the range of values. You can see the functions f(y) and g(y) as in this case simply PE for f(y) and KE for g(y) from the initial graph 1.01. The third column multiplication can be seen to have a dy/dx of being at zero at 25 for PE and KE =5. Above that point there is a negative dy/dx (or gradient) and below that point there is a positive dy/dx. The d2y/dx2 is falling at a faster rate above and rising at a faster rate below the optimum at PE+KE = 5 = PE x KE.

 This can be seen as Hamiltonian as well, as the multiplication goes on multiple points, so x= 3, y=1 is the same or similar to x=1, y=3, each one having the inner product of <x.y> = 3 multiplied by 1. Then with a lagrangian as the constraint part of the underlying structure of physics.

A point on this is that there is only one answer from all physics which is the optimum is half the value of the constraint. We loosen our thinking to allow for different phases of the parabola in graph 1.1, so there is a shifting of the parabola left and rightward, giving a large number of possibilities for prediction.

Super and sub R numbers

The preceeding goes into our theory of super and sub R numbers. Graph 1.2 shows an analysis which can be brought to bear for many difficult and intractable equations. We originally invented

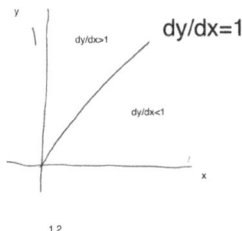

GRAPH 1.2 , SUPER AND SUB R NUMBERS

them to analyse Bessell Functions, dividing different parts of the bessell functions into graphs of super and sub R numbers.

At dy/dx =1, there is an equilibrium and we contest that R numbers in analysis only represent the dy/dx = 1 line, 45 degrees from the two axes of y and x. At here there is no movement in the variable as the change from x (dx) and the change in the function from y (dy) are equal, giving a ratio of 1. Where the equations are showing higher levels of increase it is the part of the graph that is dy/dx > 1, so there is an increasing change to the function, what we call a super R number and conversely where there is a falling rate, dy/dx<1, we call this a sub R number.

Graph 1.3 shows a parabola of multiplication, and more abstractly, the different dy/dx of the curve

GRAPH 1.3 THE PARABOLA OF MULTIPLICATION AND SUPER AND SUB R NUMBER

with equilibrium of A*. This optimum is found where the dy/dx = 0. Whilst above equilibrium dy/dx<0 and below it, dy/dx>0.

To put these into super and sub R numbers you simply look at the many triangles of gradient, which we can call *dy/dx*

At the point of equilibrium the dy/dx = 0 but the *dy/dx* is equal to 1. This can be seen as an eigenvalue (a stretching of the vector by the characteristic root). What's new with our analysis is that below the equilibrium in the graph 1.3, the *dy/dx* >1, so has a super R number (that is an increasing rate) below A*. The *dy/dx* <1 where the dy/dx<0 giving the sub R number.

Our *dy/dx* ratio is called an eigenvalue is because Ax = m x, to compute an eigenvalue, so (A-mI)x = 0. To replace Ax by dy and mx by dx, you obtain *dy=dx* for the eigenvalue in normal analysis, and super and sub where the equation is going at lower and higher rates of change. R numbers are therefore solutions to the equation, as you can draw sin waves orbiting the R number of 1, so as it

6

crosses (has R = 1) there is a solution or eigenvalue of the equation. This can be applied to partial differential equations, e.g

$$\frac{d^2x}{dy^2} + \frac{dx}{dy} + x(y)$$

Then you draw out a model of R numbers for dx to dy, the second differential of dx and dy and also the x = f(y). The benefit of this is to be able to deal with much larger orders of differentiation with where the line in the model crosses R=1 being an equilibrium since there is no more change in the variable. In addition as the line crosses one R (for second order differential) and another, you have a tractable way to reverse engineer equations and come to their eigenvalue (characteristic roots or solution or point of attraction).

Waste functions

The kinetic and potential equations can be brought to life with a waste function. Much like an economic system where nodes trade with other nodes, there is an iterative map of x:-> x (waste function, x) where a negative value added in the mapping (the trade) exists as a waste function.

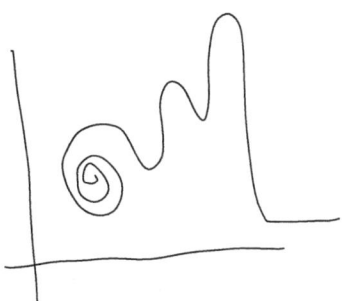

Graph 1.4 of attractor of first law of complex systems theory, the kamaa equation

The distribution of waste and positivity in mapping leads to an interesting mathematical answer for many nodes (physics often restricts itself to a single variable, or a multiplication of the variable by the number of nodes, for example a multi particle system of forces). This then follows the structure of the first theory of complex systems analysis: a triangle of nodes mapping onto each other by the waste + positivity, which is a law of complex systems since it has the same overall dynamic no matter how many nodes there are (where nodes >= 3), see graph 1.5.

Graph 1.5 First law of complex systems theory, mappings between all nodes with each other

It is possible to imagine a small radius of an attractor of the first law of complex system's theory's modelled systems, as at lower rate being a spiral and then where there is a large positive perturbation the attractor goes up and down on an increasing rate that then moves to zero. This is very hard to explain, and depends on a certain distribution of waste functions and negative waste functions (value added). Given an even level of waste's variables distribution there will be as many positive values from the complex system mapping as negatives, so we would informally expect there to be a spiraling attractor as the variable goes up for a while then down, then further in limit locally to an equilibrium. With a small perturbation there is a movement out, which when the function is followed becomes drawn in a spiral to the equilibrium. For a large positive perturbation there is a rising and falling and rising more then all together the system loses coherency and goes to zero.

As the complex system law applied to itself, in other words added onto itself, is the same triangle of nodes and mappings, then the model of this, the attractor of the first law of complex systems theory is simply the same graph and process, graph 1.4. This then can be made into a ring that always has the same structure and properties. This ring is the same complex systems law.

Looking at the mean value theorem, applied to the first law of complex system's theory, we would have :

$$\text{Lim } x = \frac{f(x+h) - f(x)}{h}$$
$$X\text{->}0$$

The f(x) is the attractor in graph 1.4, and a moving along of the function, by the perturbation, h, gives an interesting digression on Taylor Series expansions. We can respecify Taylor series expansions as :

$$\frac{f(x - h) + f'(x-h) + f''(x-h)}{n/2^n. \quad n/2^n} \qquad \text{Equation 1.6, taylor series expansion}$$

Instead of factorials that dampen the effect of higher orders of differentials, we have the series n divided by 2 to the power n. This has the same effect of lessening the result of each differential in the expansion. The h is the guess of the value for optimum, and it means only a local equilibrium, i.e. it only is correct in a portion of the function being optimised by the Taylor Series. In a sense you are doing the same thing as high school math where a line of $2x + 3$ is found from a line where you find the gradient to give the coefficient with the variable x, while 3 is found by the crossing of the y axis by the function. This linearisation is simply added on in Taylor series expansions. What we assert is that the Taylor series expansion equation 1.6, is simply a following of the function from the initial guess for the variable, h (because the function is simulated as the difference between h and x.)

Taking h as a perturbation we see that in most probability the attractor will go back to the fixed point at the centre of the spiral from graph 1.4. With less probability (where h is large or persistent in 1 direction) then the equilibrium is zero at the trajectory of the attractor at the two humps and falling away to zero.

Proof:
Theorem 1.0

A Taylor series expansion is defined as:

f(x) = f(a) + f'(x-a)

If x-a >1 then a increases by the change of function f'()
If x-a < 1 then a falls by the change in the function f'()

Putting on the factorial dampening, you get f''(a)/2! + f'''(a)/3!, etc
This dampening is assumed to have little effect for many equations (where there is little movement in the variable from f''(a) and upwards in order of differentiation). Thus they can be ignored for the Taylor series expansion.

Since x-a >1 or x-a<1 the repetition of Taylor series leads to a local optimum and the result following (calculating each step) the result from the function. It is at the optimum that x=a, since if it was above it would be less and if it is below it will be more.

What we mean is that the Taylor series expansion is put into play with the guess, a, then the expansion follows the path of the function being expanded (f'(x)).

This concludes the proof.

<u>An overlooked part of eigenvalues</u>

The eigenvalue is the solution and determines the dynamics of a system of equations.

However, what is often overlooked in the literature is the solution found of the eignevalue's characteristic equation is that it depends on a dominant eigenvalue to make a solution.

There can be many eigenvalues which we look at as coming to one value $\lambda*$ if they are in symmetry and balance with each other.

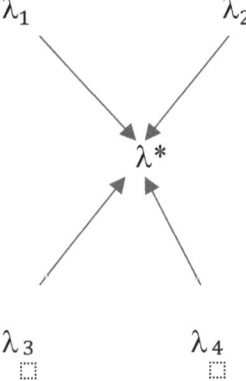

Figure 1.6 – eigenvalue system, $\lambda*$ is dominant eigenvalue, λ_i are other eigenvalues

The above example of eigenvalues of a system of equations, figure 1.6, has the eigenvalue $\lambda*$ if the other eigenvalues, λ_i are equal to each other. Consider though there being more push from one λ_i than another, then there is a dynamic of change in the equation system. Each λ_i is a point of attraction, like a magnet drawing the plot of the variable. The ordering of the eigenvalues is very important in this and shows that there are many possible trajectories of the variable.

Section 3

Looking at the Kamaa equation (see graph 1.4) you could very easily see this as better analysed as Fourier transforms.

This leads us to consider deducing Fourier Transforms from Euler's equations.

Euler's equations reduce dimensionality in an analysis, given that they are $\cos(e^x)$ in the real axis and i sin (e^x) in the imaginary axis. The drawing of a circle where these are two radii where the radius is e^x gives a connection between sin, cos and exponential. Then an equation is simply put as the vertical height of a cylinder where the Euler equations are a cross section of it.

This is necessary to solve the Fourier transform, and coupled with our pont de capiton of Taylor series, it becomes clear that there is a problem.

Thus we lead the reader to a novel interpretations and algorithmic structure for solving these very hard equations.

To give our investigations colour, we take Fermat's last theorem and show a proof through the concept of the crown and the labyrinth.

Consider the story of the Minotaur in a maze. The person going into the maze puts a line of string along their path through the maze to find the Minotaur. When they have found it they simply follow the path back shown by the string they have laid along their path. This leads you out of the maze and the exit/entrance is called by us the crown.

Mathematically, you could have a Jacobian matrix of Fermat's last theorem, $a^n + b^n = c^n$, rearranged into three equations, so there is $a^n = c^n - b^n$ and also $b^n = c^n - a^n$.

To solve this, you can follow all sorts of different lines (we believe that all is flux from antiquity). So a rise in a leads to b rising, then c rises, though we could explore and simulate the perturbations between those equations in many ways.

We have the concept of the crown. Where you simply solve the equations for one variable and notice the change in different values in this system of interdependent equations, in a sense pulling the thread from the maze of simulated values in this system by choosing one crown, say all from a^n changing other values.

Ordinarily you would solve this kind of equation system with Fourier analysis. This leads to fluctuations and oscillations that are endemic in this. Since Euler's equations, at least in our interpretation, are unnecessary simplifications, this would lead us to look for this approach of crown and thread in the labyrinth to solve the equation of Fermat's last theorem.

The associativity, distributive and commutativity laws as well as cycling prove the crown and thread algorithm.

Associativity: $a + (b + c) = (a + b) + c$
Commutativity: $a + b = b + a$
            $a\,b = ba$

Distributive law : $a (b + c) = a\,b + a\,c$

Cycling: (from first law of complex systems, ordering does not matter for relevant systems)
            $a \to b \to c$
            $c \to b \to a$
            $b \to c \to a$

The proof of this is left to the reader, as it is fairly trivial. This proof is the answer to Fermat's Last Theorem. Clue: look at the effect of symmetry in assumption and structure and symmetry in process of the thought.

The interesting advance is this leads us to consider the distribution of a Jacobian matrix, especially for large numbers of variables, so there is a surface drawn along each element of the matrix. A normal distribution would mean elements in the centre and near this have higher values, thus there is more stability, at first blush, while more extreme elements (e.g dx/dx) having less effect. Skewed Jacobian matrices would have different dynamics. We leave it as a point for future research to analyse and simulate the flows for different distributions of Jacobians.

Section 4

From Jacobian to Ito's formula

Ito's formula and deficiencies (martingale is heads or tail)

Ito's formula comes from an analysis of martingales which are from the unit of analysis being the flip of a coin into heads or tails. From this comes the martingale theory, that the expectation of the variable is zero, because on average the amount of heads from a coin flipping is the same as the amount of tails that come about. The martingale is like the atom or molecule in physic's analysis. This is the main deficiency of Ito's formula, that it only models a random system of true or false, 1 or 0, head's or tails of a coin flip. Considering the much more complex theory of stock market investment, there is a limitation in simulation (fitting data to the martingale and also modelling something that does not increase the amount of information in it) and invention (what to invest in). This as our Kantian leads us to a new model.

We will show you an alternative model from what is called Zain's paradox. This is an inference from the theory of $1 + 1 = 11$. Zain's paradox is needed to be a unit of analysis in 'free math' for everything. Essentially fields are split and then recast together. So the Jacobian distribution of the stock prices against each other are split (for example possibly Frobenius splitting of fields) then put together as a portfolio of shares. From the Jacobian distributions this can be seen as a better way to forecast stock price returns than Ito's Lemma and stochastic calculus. See models by successful market practitioners of stock investment (the heat diffusion equation, which can be rewritten using this method).

11

Matlab simulation of min tayn thumma

The circular spiral of a kamaa equation is identifiable as the min thumma shape, a tangent bundle of 3 or more circular paths of a vector field around min-thumma.

Figure 2.0- min thumma,

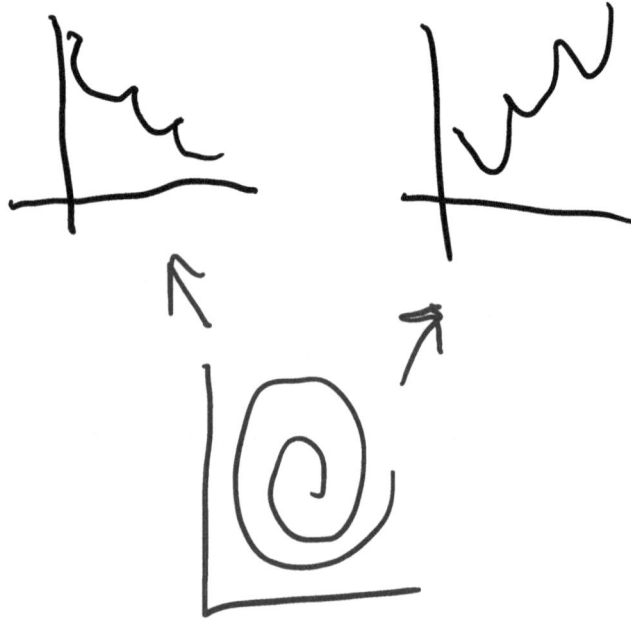

The spiral of kamma equations can be reduced in dimensions to the min-thumma curves. The line of the equations and graphs are on the line, a separate boundary, a clay that joins together the different parts of the whole space (a sploge- where every point is set at a positive number, is the opposite of this). As the clay, the lines in above graphs, is a separate set from both sides of the graph's space, there is an easier to demonstrate trajectory and dynamic.

This is an advance in the underlying theory of physics, the oscillation from opposite functions that leads to an equilibrium, but in min-thumma you involute the parabola in each underlying function of the system of equations.

The division of time book I, mathematics and the interdependent revolution

0.0 Information

1. We take a preference operator of High > Low levels of information in a piece of mathematics. A function can be changed round to have less information or be operated on and have more information. This is the core of mathematics.

2. Using e^x or Euler's equation, inner products, fourier analysis or trigonometric ratios reduces and spends information and dimensiality of the equation be investigated.

3. The commutative law is assumed and later is sometimes relaxed, so A ∩ B = B ∩ A, so Ab= bA. We shall see that this law being relaxed leads to nonlinear dynamic properties, being a feedback in a variable of a function in the function itself.

4. The distributive law, where x(y+z) = xy + xz

5. The associative law, where (xy)z=x(yz)

6. The four witnesses of tahir's box, discussed in other work by us. Consider the division of a rectangle into two parts, one from the horizontal framing of the rectangle and one from the vertical framing of it.

7. Division: AB=BA, but
$$\frac{A}{B} \text{ does not equal } \frac{B}{A}$$

But in two dimensions you could have a 3 by 1 rectangle and a 1 by 3 units rectangle. Thus simply defining a relative sign of variables does not hold all the information necessary to operate and deduce from them, indeed also this leads to them having the same area, yet are different divisions of the space.

   7.1 integration by parts holds from addition in the assumption being distributive. Yet with the point in (7) about two dimensionality and mathematics, you get two results when you draw this out as a rectangle. We leave the proof of this trivial point to the reader.

8.0 Dimensionality is crucial in our analysis, a line is one dimension, yet a square is two dimensions and in three we see the labyrinth (the endless combinations of different equations, functions, variables, matrices, means of operating on matrices) as a triangular quarter pyramid, a cube, a cone a sphere.

9.0 Kammaa equations (from our other work) can be represented as a cone for the initial spiral of the phase space of kammaa equations, a cube for the w-shaped plots and a quarter pyramid as the collapse to a lower level than the initial spiral.

10.0 A conservation principal makes the geodesic, the shortest path, leading in analysis to the triangle inequality and also therefore the derivative dy/dx. But consider movements of triangles inside a sphere as a surrounding basis!

1.0.2 Exponential and Euler's formula

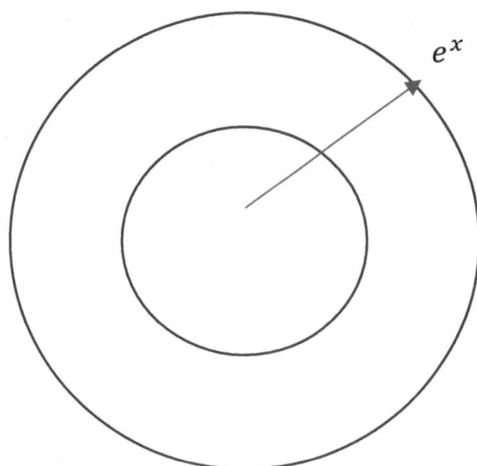

Figure 1.0.2 Euler's equation and exponential as a radius

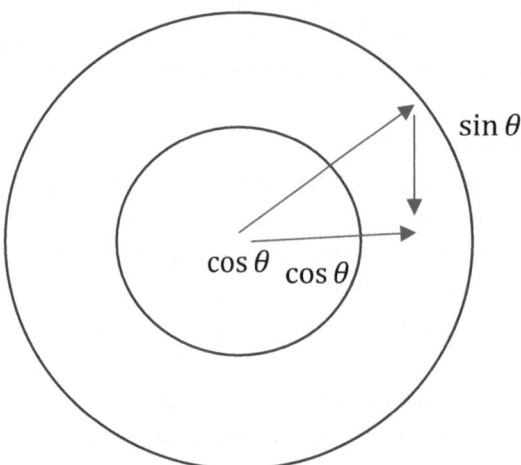

Figure 1.0.3 Trigonometric ratios and exponential e^x

From the figures 1.0.2 and 1.03 you can see that the sum of squared sides of cos and sin θ is equal to the square of the radius, e^x. We argue that this leads to Fourier analysis as an arbitrary radius for the circle and its ratios every 90 degrees in rotation of triangles like the above. Answers are given, yet we feel somehow cheated by the mathematics. For example you could have any radius for the circle of cos and sine which as it goes higher in θ that have the period of 2π radians in full circle and $\frac{\pi}{2}$ radians for the functions of sine and cos. Yet in fourier analysis you use the Euler's equation radius as e^x. The introduction of e^x leads to a flattening and loss of information with its property of being the same if it is differentiated.

Exercise 1.0.2

Determine the dialectic (the opposite) of e^x
What is it's inverse?
What is not e^x? (in logic)

2.2 Bessel Functions

Bessel functions are combinations of sin and cos, placed together, giving the feedback of vibrations on the head of a drum. We can see this knits with fourier analysis given the above relations in figure 1.0.2 and figure 1.0.3.

One thing is clear, exponentials lead to oscillations if put against each other and with different values and functions.

2.4 Partial derivatives and exponentials

A system of equations of partial derivatives can be expressed as:

$$x_i + y_i + z_i = 0$$

A Jacobian (much related to a covariance matrix in statistics) of this equation's partial derivatives is:

$$\begin{pmatrix} \frac{\partial x}{\partial x} & \frac{\partial x}{\partial y} & \frac{\partial x}{\partial z} \\ \frac{\partial y}{\partial x} & \frac{\partial y}{\partial y} & \frac{\partial y}{\partial z} \\ \frac{\partial z}{\partial x} & \frac{\partial z}{\partial y} & \frac{\partial z}{\partial z} \end{pmatrix} = J$$

Given a certain range of values for each element of the jacobian J matrix, it leads us to a series of weights for analysis in lic groups;

$\delta > 1$ then $e^{\delta t}$

$\delta < 1$ then $e^{-\delta t}$

$\delta = 0$ then 0

Taking the δ as a line then the mean between them gives an equilibrium or perhaps eigenvalues/characteristic roots/ zeros, as $e^{-\delta t/2}$.
Asymptotically each one leads to an infinite value (in absolute terms). In other words the limit of lie group weights as $e^{\delta t}$ and $e^{\delta t}$ as delta tends towards infinity is the same, infinity. However for say $e^{\delta t}e^{yt}$ the limit is the same of infinity but the level of tightness of the spiral in kammaa equations (in other words the solution to lie groups from Jacobian matrices) is different, so some phase spaces are high in change or lower given the same perturbation.

5.0 Norms, metrics, closed sets, concavity, inner product, gradients, limits/differentiation, basis, pythagorous, triangle inequality

1. A norm is a metric of a space. You can see it as the geodesic (central to all mathematics) of a set of points or lines that form a closed set (always have numbers between any two points in the set) and is also only possible in a closed set.
2. Hilbert spaces from quantum mechanics use inner products to define the metric of those quantum.
3. Gradients of curves with limits rely on the norm and closedness of the set being convex.
4. Gradients of vectors are more dubious in the fundamental axioms required to define a norm / metric.
5. Gradients of a curve rely on division and form a group. There is the problem of division as can be seen from the prime divisor theorem.
6. Gradients of vectors use only one line from the jacobian matrix (J) and so do not carry on the interdependent dynamics, which are in face remarkably simple (from lie group weights).

Kammaa equations derivation from exponential weyl lie group weights

Section 1

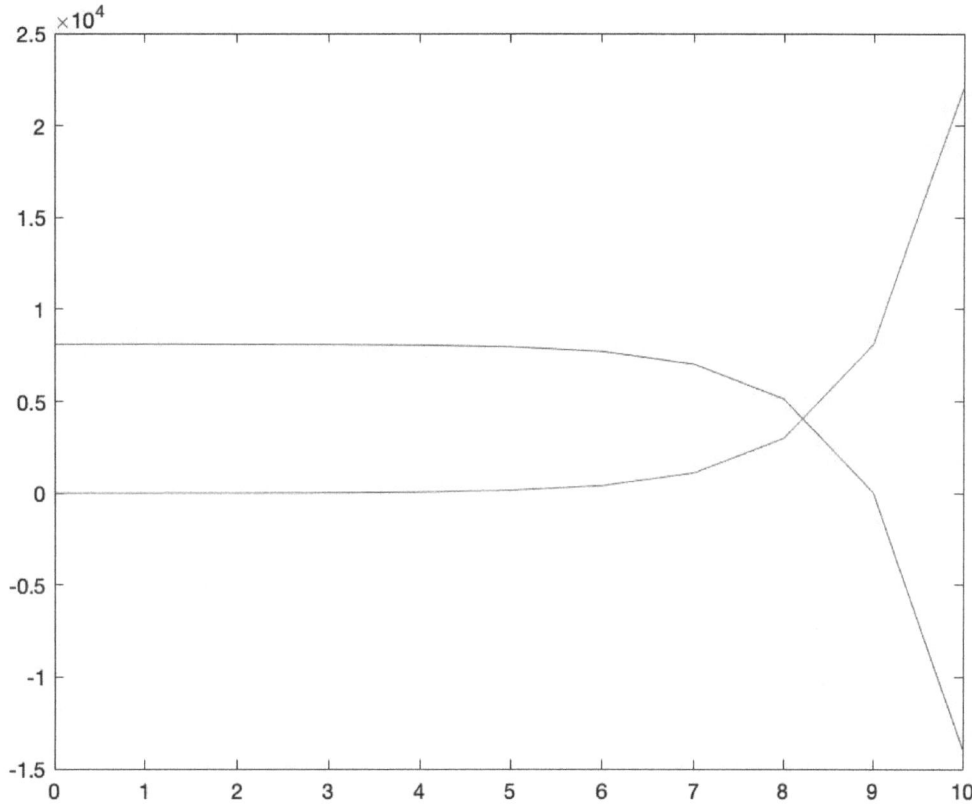

Line sloping upwards is e(x), line sloping down is 1-e(x)+shifting parameter (highest number of the e(x) line). These are weights of lie group taken to time is equal to infinity. This is an abstract structure. Yet it is illuminating that you can look at all kinds of random data, such a stock price returns, and see and analyse into different parts of the kamaa equation; a spiral and 'w' shaped time series.

Difference between lines given in graph 2. At the crossing of the e(x) and 1-e(x) lines, there is the bifurcation of the time series. It is abstract in that the shifting parameter to make and determine when the time series bifurcates is arbitrary.

Graph 2.0 the bifurcation of kammaa (spiral below zero and w shaped rising then falling to zero after asymptote of the exponential.

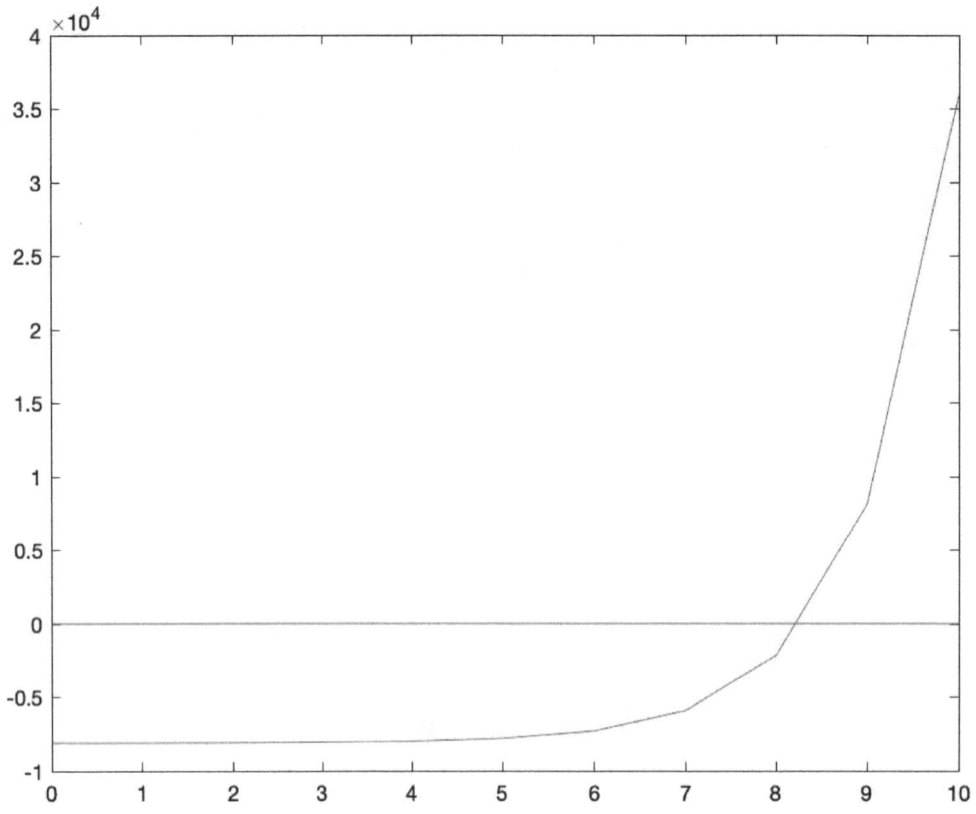

Exercise

Then do as a Hamiltonian field whose basis is a sphere,

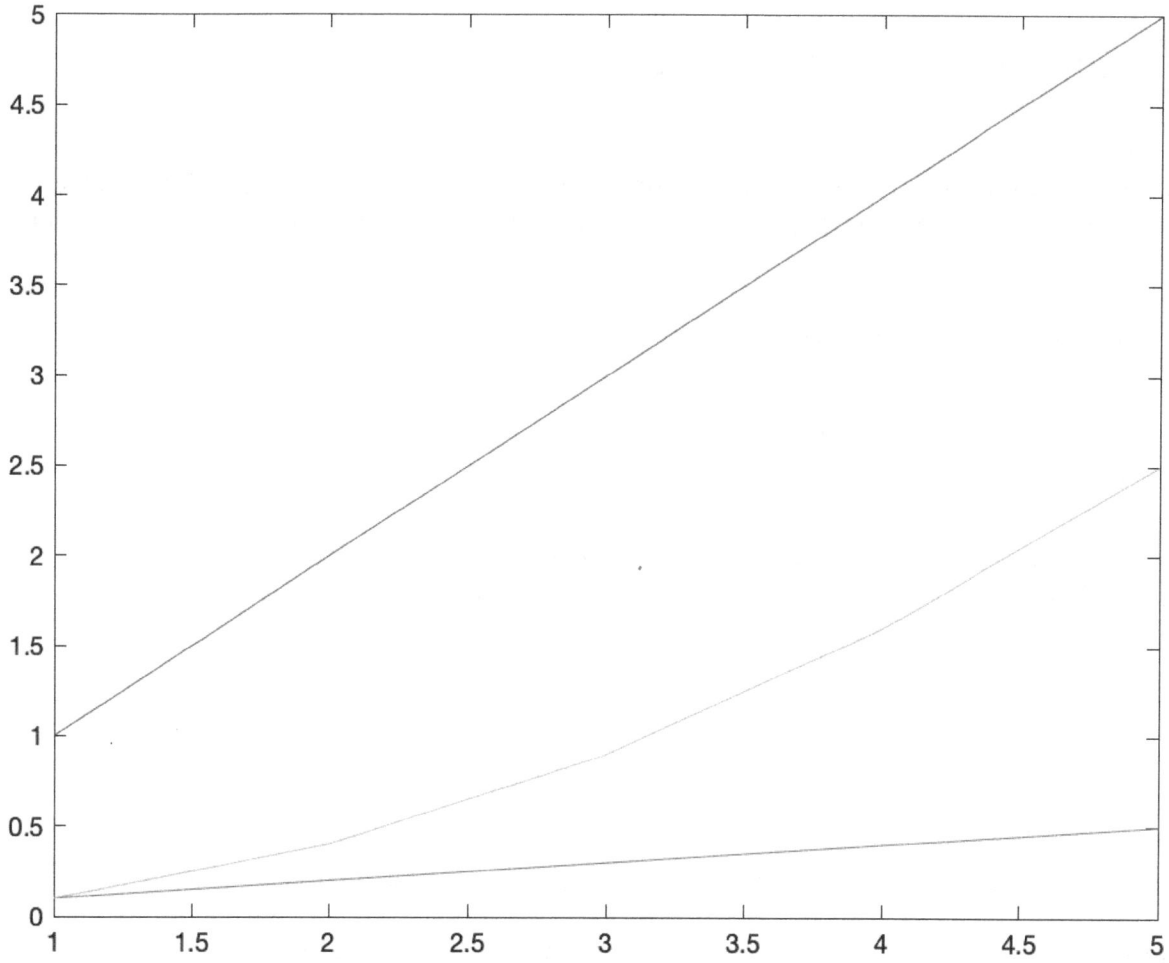

Above graph 3.0 is p (in line at top of graph), the Hamiltonian (h) next in the graph and at the bottom line ,q .

We generate a sphere for this , seen as the basis of the Hamiltonian. The proof of this is integration by parts, as each square of the plots of the data, q and p, are summed together to represent a tangent plane that goes round into becoming a cylinder around each part of the sphere of basis (closely linked to the metric of the Hamiltonian sphere). As given above, the squares can be different in size, for example a rectangle or a larger and then smaller square or rectangle giving an elliptoid underlying surface to the Hamiltonian.

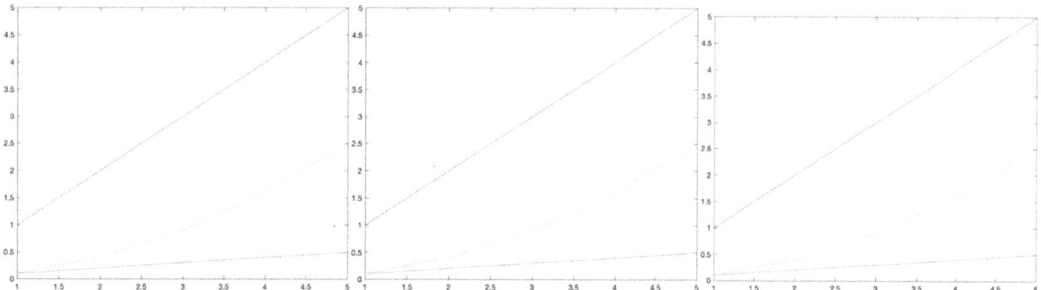

Graph 4.0, each of the cylinders wrapping round the Hamiltonian basis sphere, so at initial condition, the first differentials with respect to time for p and q and h as p multiplied by q.

Integration by parts is:

$$\int p'(t)q(t)dt + \int p(t)q'(t)dt$$

Then differentiate this to obtain p'(t) and q'(t) for the different graphs 4.0 as a sum of the product area, p multiplied by q. So we can approximate the functions by differentiation of the variables and then summing them.

The Hamiltonian (H) can be seen as:

$$\frac{dp}{dt} \cdot \frac{dq}{dt} \cdot \frac{dpq}{dt} = \partial H$$

This then it applied to a sphere basis as above in graph 4.0 and with integration by parts then differentiated to give the result as a sum of the different cylinder sphere squares of graph 4.0 , so the taylor expansion of a sum of p to q relations is:

$$p + q + p_2 + q_2 = \Psi(n)$$

Where $\Psi(n)$ is the taylor expansion and n is the number of cylinders looked into, so here n=2. You can see this solution from limit theorems of calculus (as the division of the top is 2).

In addition you can look at a series such as (1,2,3,4,5) and find a kind of differentiation at each interval of the integers here.

Series: 1   2   3   4   5
Differentiation: 2-1=1, 3-2=1, 4-3=1,5-4=1
(in other words it goes up by 1 for each element of the set)

This is used in our simplified taylor expansion above for $\Psi(n)$ and assumes in this case a 1 unit dx of dy/dx.

Now the process of making answers to a swathe of unsolvable equations has begun, and in the many new technologies from these theories could bring a new era in advancement for the world.

Exercise 1.0

Do Taylor expansion of basic Hamiltonian sphere basis/metric. This is the special theory of lie group weights leading to kammaa equations. What is the general theory for non-homogenous spheres, for example where the elliptoid basis/metric and morphing shape in expansion and contraction (see below Lam and Dhal, section 1.3) .

Exercise 1.1

Then do Hamiltonian of bifurcating kammaa equation : two sets of sphere's cylinder squares; for e^x > $(1-e^x)$ and for $e^x < (1-e^x)$+shifting parameter, shifting parameter = upper domain bound for e^x.

Exercise 1.2

Then do Taylor expansion based on the bifurcating kammaa equation (should be two different Taylor series expansions- two squares of the sphere's cylinders unwrapped).

Section 1.2 Ba-a-na equation

This is two parabolas defined as the underlying theory of physics's Shylla and Charibus (symmetry and balance) of each variable being added together to be a constant and then the index of these variables multiplied by each other to give the vertical axis in the ba-a-na equation given in graph 4.0.

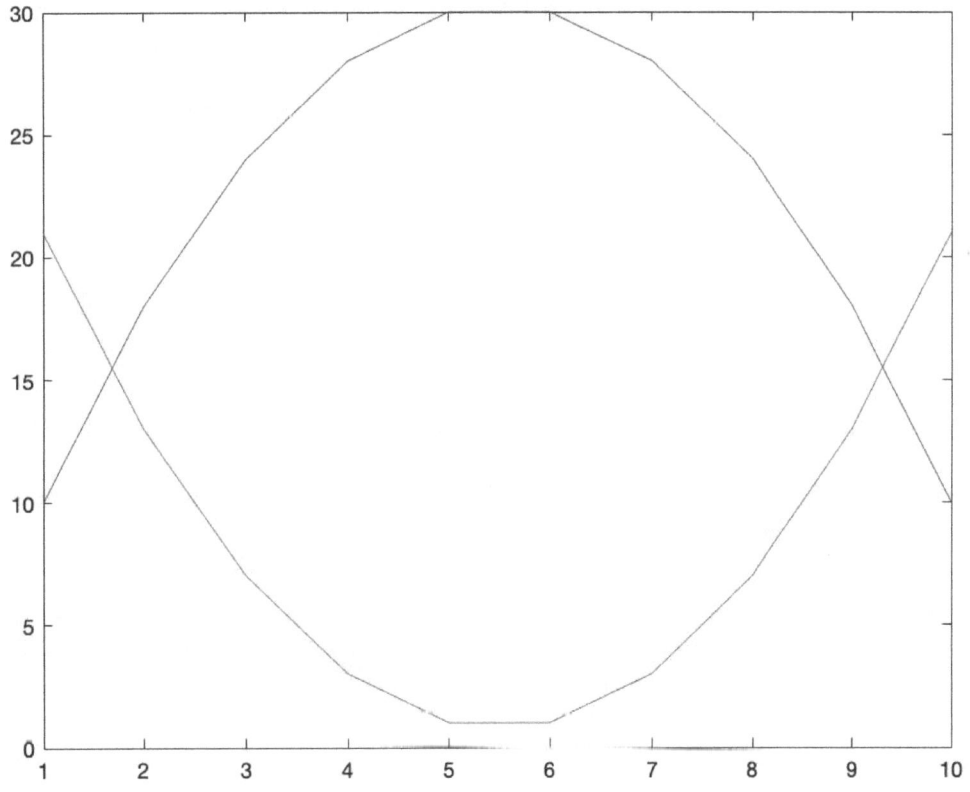

Graph 5.0 – ba-a-na equation generator from x+y=constant, then graphing x times y for the top parabola, and 1-x times y + the constant for the opposite parabola. They cross because of the shifting parameter (in graph 5.0 this is 30), the constant here that makes an interesting core generator of phase space, perhaps you can see it as the mathematical basis of an attractor.

The key to interpreting this graph is that the difference between each of the parabolas leads to a modelling of kammaa equation spirals, where the centre is the radius of each loop of the spiral going form 2 to 9 in the horizontal axis, while the limit of the differences between the parabolas for less than 2 or more than 9 , are increasing, so the w-shaped part of the kamma that then ends where the graph is not shown and overloads or becomes zero, depending on which side we are looking at in graph 5.0.

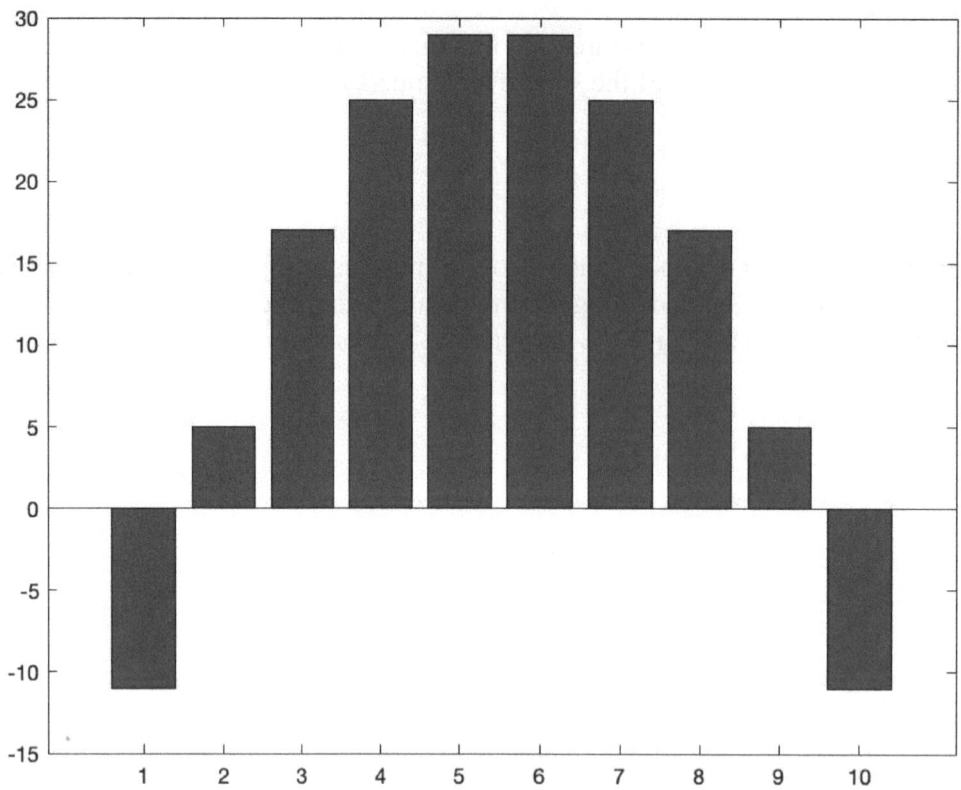

Graph 5.1- the distribution of values of each radius in the difference of lines in figure 5.0.

The negative areas mark the falling out of the variable's phase space trajectory, where they reach either zero or infinity after leaving the phase space of graph 5.1.

The positive areas in the middle show a probability distribution which can be interpreted as a wave function in quantum mechanics, where the quantum is most likely to be. This suggests a negative space (more than shadow in art, the dialectic of the quantum field) where connection is possible with positive quantum wave functions superpositions. This leads to the invention of quantum switches, for example in computer memory, allowing huge amounts of data to be stored in a small microchip of RAM.

Exercise 2.0

Write a superposition of quantum wave functions using the above work (add it up or the opposite! clue). How does negative space add up to a force of bonding and clustering of quanta?

<u>Section 1.3, Expanding and contracting (Lam and dhal)</u>

An important lemma in our analysis is that of contracting and expanding functions numerically and in magnitude.

Lam operations are expanding of the function/ phase space diagram, multiplying a phase space by a constant or function about one variable.

Dhal operations are contracting the function/phase space diagram, dividing the phase space by a constant or function about one variable.

What is important is that if a single variable is expanded or contracted then the ratio of the domain of the function is reducing / increasing by a higher gradient or difference or differential or lower gradient/difference/differential.

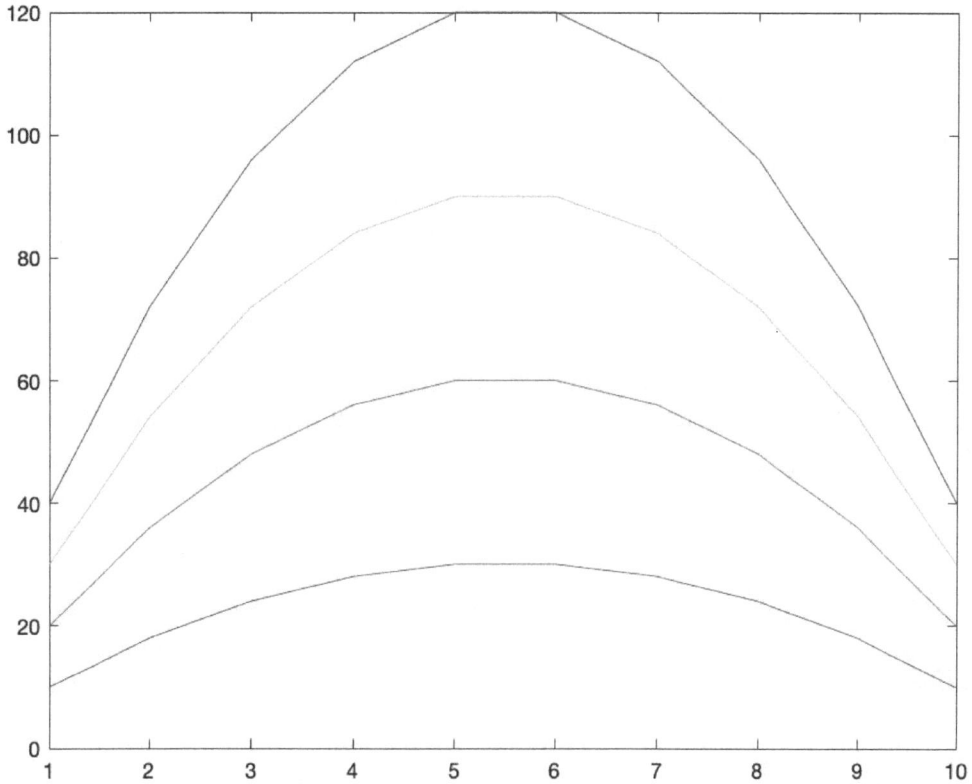

Graph 6.0 the expanding (Lam) and contracting (Dhal) of a parabola, each one a multiplication of x 2, x 3, x4 of the original function, notice the change in differential of the parabolas. This can be given as the rising in differences of each parabola as you go up the vertical axis or central part.

Thus lam operations increase the differential of the curve, whilst dhal operations reduce the differential of the curve.

Proof: that rising parabolas (ba's) have higher derivatives.

Assume a function is a parabola = x_1

$$dx = 1$$

For each parabola with lam operators of 1,2,3,etc:

$$dy = 1, dy = 2$$

Then the gradient of the parabola (the derivative) is:

$$\frac{dy}{dx} = \frac{1}{1} = 1$$

For dy=2, in other words a twice as high lam operator expansion, its gradient is:

$$\frac{dy}{dx} = \frac{2}{1} = 2$$

So the gradient is higher with x being fixed and the dependent variable going double. This is really quite clear but when applied to interdependent equations and jacobian matrices it has a huge effect on the process from such a structure. This concludes the proof.

A short digression on quantum wave functions and fields as a ring

For the distribution of a quantum, the wave function, if added together make a superposition. However, as a ring of multiplication and addition, there is a rising gradient for a given dot in the quantised field.

Exercise 3.0

How is a ring of addition related and the same as a ring of multiplication?

Ring of wave function superposition

As we take the underlying generator (the structure) as two parabolas of ba-a-na equations, then multiplying each parabola makes a higher differential of the superposition, in this case the superposition found through multiplying the parabolas, which as we see is then graph 6.0's rising parabolas. This can be shown through induction that there is a superposition (layering of quanta) as a multiplication, which needs the concept of feedback or in other words interdependent equations.

We can show interdependent equations as a kind of jacobian matrix:

$$\frac{\partial x}{\partial x}, \frac{\partial y}{\partial x}, \frac{\partial z}{\partial x} \begin{pmatrix} 1 & 0 & 0 \\ 0 & 1 & 0 \\ 0 & 0 & 1 \end{pmatrix}$$

This is for a very simple system of equations where x+y+z for the changes in variables, x,y,z.

$$\frac{\partial x}{\partial x}, \frac{\partial y}{\partial x}, \frac{\partial z}{\partial x} \begin{pmatrix} 1 & 2 & 0 \\ 2 & 1 & 3 \\ 0 & 3 & 1 \end{pmatrix}$$

This next jacobian matrix is for x+2y+0z, 2y+ x +3z and 3y+z.

As you can see the equations given here are interdependent, so they feed into each other. A rise in x leads to a rise in z then more x and y, then into the original x. Heraclitus of Greece made the same concept in his work, the Fragments, on the relation between fire, water, earth and air.

Note this is different to a normal Jacobian matrix which shows all the different differentials for each variable, here we are expanding using x , (multiplying by x) as the derivative of the function goes higher with higher lam of x (expanding of x). The ratios of the curves are rising then as x is higher / lam operation.

We can now do a superposition by changing the matrix above that are multiplied by derivatives. The multiplication is not symmetric as much of maths uses, but rather makes different shapes and different bulging of the lam operation resultant shape.

Exercise 4.0

What is the trajectory from above or below the ba-a-na equations of graph 5.0? Clue, the answer is just above and below is an equilibrium. Note how to prove this analytically?

Lorenz equations, a curious example of interdependent equations

The following matlab code shows a problem in determining interdependency in Lorenz's equations for the weather.

MATLAB Lorenz equation code:

```
a=2;
tdx1=[0,0,0,0,0,0,0,0,0,0];
tdx2=[0,0,0,0,0,0,0,0,0,0];
tdx3=[0,0,0,0,0,0,0,0,0,0];
dx=0;
dy=0;
dz=0;

dxnew=0;

x=1;
y=2;
z=5;

sigma = 10;
rho = 28;
beta = 8/3;

%the problem of interdependent equations, the following code
%shows the lorenz equations individually plotting the data
%but it does not take into account interractions.  thus
%there is the problem of systems of equations being solved.

for i=1:10

dx=dx+(sigma*(y-z));
tdx1(i)=dx;

end

for i=1:10
dy = dy+x*(rho-z)-y;
tdx2(i)=dy;

end

for i=1:10
dz = dx*dy-beta*dz;

tdx3(i)=dz;

end
```

```
plot3(tdx1,tdx2,tdx3)
```

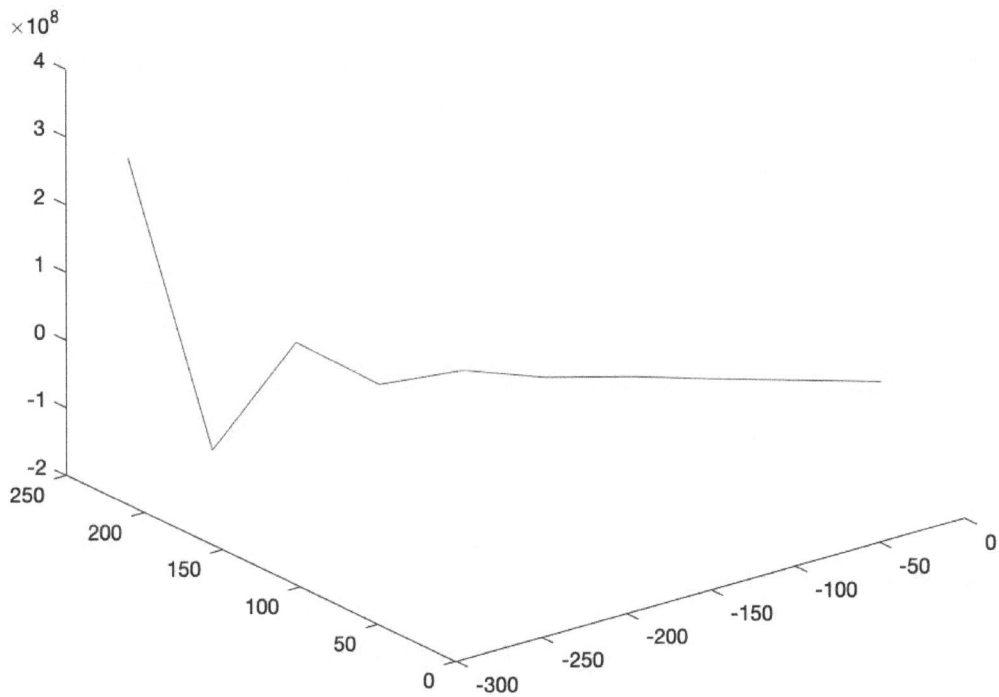

Graph 7.0 of Lorenz equations generated from above code in matlab.

As you can see the phase space in graph 7.0 is just one of many results from simulating the Lorenz equations in different orders and interactions of equations, for example, a perturbation of p leads the x variable to change which then affects the z and y equations at different amounts given the Lorenz equations which then affects x and then maybe x itself and then z. This is the problem of estimating and simulating interdependent equations. It also shows the deficiency of Jacobians in their present formulation, dy/dx rather than dydx/dzdxdy which then goes into dy/dx again.

We could look at network and graph theories for this, yet the present state of this is also leaving the thirst for knowledge wanting.

Exercise

How do you make a jacobian matrix of the Lorenz equations? Clue, at its base is actually symmetry and balance, the countervailing forces, like torque and torsion, making a cancellation of each other, yet also a special theory of oscillations given by kamma equations. Especially there is a secondary inversion of the kamma spiral in this equation.

Exercise

Consider the statistical estimation of kamma. Use a GARCH variances OLS on data or in simulation to analyse kamma equations.

1 Generalising OLS regression for 3 or more variables (GLS)

The solution to a single variable OLS regression is:

$E(X|Y)$

For variables Y causing X. The expectation here is the gradient of the fitted line of the data, often calculated by minimising the data's differences from the mean or average.

For more variables, then the OLS regression breaks down because there is interdependency between variables that finding differences from mean does not work for.

So for this kind of OLS regression we have the following probability measures expectations:

$E(X|Y, Z)$
$E(Y|X,Z)$
$E(Z|X,Y)$

Clearly you can see through the labyrinth's thread and crown (see section 3 before), that the variable in changing has a feedback between the expectations, thus making it difficult to solve and put numbers to in real life, without a new theory for solving OLS in 3 or more variables.

In addition to this, there is the potential to parametrise kamma equations, which are the phase space of interdependent equations and their systems.

In OLS regression, this problem is sometimes solved with a matrix repetition of inversion:

$C \times C^{-1}$

Where a matrix is multiplied by a matrix and then the inverse (in others words going backwards) is multiplied by it. Given that $C \times C^{-1}$ does not have hysteresis (the compounding of time and change of the variable) then this would seem possible. The information not put into this equation $C \times C^{-1}$, is what is the expectation, what is the distribution and what is the covariance, before looking at asymptotical analysis (what the functions become over long periods of time).

1.1 Why is there hysteresis

Because of interactions there is hysteresis. For example we can look at the cellular automata as a strip of values that have a rule to change which values are on and which are off. Say this strip for the ith run of the cellular automata is x_i , then we can through the axiom of additivity (that sums can be expanded into the combination of sums they represent) say:

$$\sum x_i = \sum x_j + \sum x_m \qquad \text{(where i= j+m, thereby splitting the sum)}$$

But with lie interactions (the interdependent equation) added to this, then an additional part must be added to the expansion, so the sum of the product of x(j) and x(m) is part of this. Simulation of this can show that there is no way to go forwards and backwards to achieve the same progression (in other words values of x(i) for all I, the full running of cellular automata).

Thus doing an inverse , such as $C \times C^{-1}$ does not uniquely specify an answer, the going forwards of C and the going backwards of C^{-1} has hysteresis because of interactions (x(m) multiplied by x(j)).

2.0 Fourier, a competitor to lie group weights

We look at a few Fourier analysis applications and discuss them; vibrations in a string, heat convection in solids and sound/signal processing.

2.1 vibration of a string

A string can be modelled as a sin wave, with a frequency and time variable. Over time however we can define a function of the frequency and time, where frequency 1 is lower than frequency 2, as

$$f_2 = f_1 + x$$

The function is x. which is related to the time variable (how long you observe the string), makes an interdependent equation so solves due to lie groups weights, to e^x

2.2 Heat conduction in a solid

We can define ϑ as the temperature in a solid, with an analysis of a cylinder that is hot, whilst the box around it is cold. Why a cylinder? We argue that this is the basis, the essence of Fourier transforms.

The Fourier transform makes a function into a disc, with an integral also of the length of the cylinder of all the discs multiplied by the disc.

Exercise 2.2

Show all Fourier transforms can be interpreted as discs and towers of discs to make a cylinder. Clue, consider the point of a varying angle of a right angled triangle in a circle, thus cosine and sine become a circle). Also what does the division by 2π make in the equation's systems of thread and crown from the labarynth. This is the argument, but is for you to decide if it is right.

Exercise 2.3

Consider the Fourier transform:

$\int_{-\pi}^{\pi} e^{-x} \cos(x)\, dx$

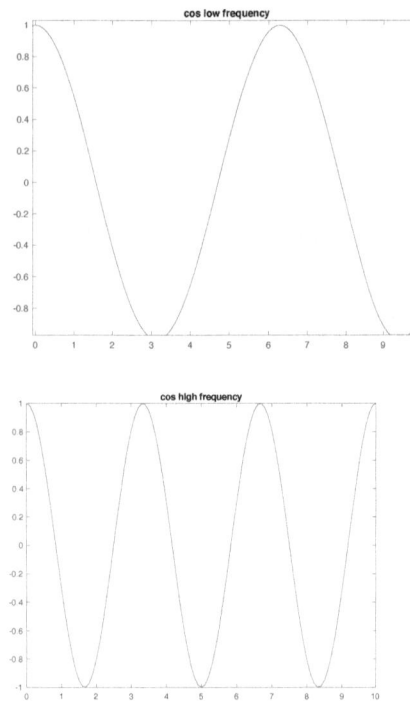

Graph out the e^{-x} function, then using thread and crown analysis of labyrinths, show the path and trajectory for the two different frequencies shown in the figure above. In MATLAB, you graph and analyse the transform by the command x= cos(0.02 * pi* (frequency) * time).

2.3 Fourier analysis and signal processing (sound engineering).

Fourier is a way of expressing sound waves, of frequency (pitch, high or low tunes) and volume. Since it is as its basis, cosines and sines, applied to exponentials and cylinders of sound, there is a beauty in the kind of sounds and usefulness of the range of sounds you can use.

Consider though, stripping out noise from a sound or signal. With Fourier analysis you simply go through the process you should have derived from exercise 2.3 on the signal. This makes due to hysteresis a sound artifact, rather than a unique result.

2.3.1 Cellular automata and noise stripping out

A set x_i is a matrix of a cellular automata (in a kind of game of life), where each I is a run of the cellular automata (which we call CA).

The rule of the CA is:

$$x_i \rightarrow x_i'$$

So the result of the rule is x_i' for all (i) from $\mathbb{R}$.

Theorem: There exists a unique solution to the CA set in x_i.

Why? x_i exists, therefore x_i' exists, as both are in $\mathbb{R}$.

Not too interesting? We now see the fruit of our labours of Hercules to challenge and defeat the truth. For many dimensions, in other words many variables we have:

$$\begin{pmatrix} x \\ y \end{pmatrix} \rightarrow x_i'$$

But also

$$\begin{pmatrix} x \\ y \end{pmatrix} \rightarrow \begin{pmatrix} x' \\ y' \end{pmatrix}$$

And :

$$\begin{pmatrix} x \\ y \end{pmatrix} \rightarrow y'$$

But for 4 dimensions you have:

$$C = \begin{pmatrix} x \\ y \\ z \\ a \end{pmatrix}$$

Exercise 2.4

Show the labyrinth and thread joining the 2 dimensional analysis above to the 4 dimensional.

Theorem of jumping and hysteresis in cellular automata rules

Consider a rule in the CA game of life, but instead of cells directly adjacent to each other being the subject of the rule to make close cells change in value, consider the concept of jumping, so cell 10 causes cell 14 and also cell 3, for example a rule of $x(i) + 4$, $x(i) - 2$, $x(i) + 20$, for all or some cells in $x(i)$ in full running.

How do you solve this? We can set an expectation of zero for the CA, so it is quite easy to see that the solution is simply the same as the vector or variable but in minusing, for example;

$$x_i \rightarrow x_i' - x_i = 0 \qquad \text{Equation 2.3.2}$$

Thus we see this equation 2.3.2 as a key concept in solving jumping but also are a powerful tool to solve the things fourier analysis purports to find answers to.

For a function of noise as an iterative map, with a random variable, W (with a given distribution and random process) you simply subtract the iterative map of noise from the signal. This allows a modelling with hysteresis which we have argued is the weakness in Fourier analysis as well as OLS regression. For example, fourier analysis is often for a set interval between 0 and 1 of time, the result calculated for that only, rather than the full set of interdependent lags and echos (in other hysteresis).

Application of interdependent equations to economics

We have studied work from MBA (masters of business administration) courses from Harvard which teach and organise the market through the learning and application of Michael Porter's five forces model, which through many business' and corporations hiring Harvard MBA alumni, creates a critical mass for a meimetic system to dominate in what is essentially the legal creation of oligopolies. The remainder of the market is predominately in the theory of monopolistic competition (from Joan Robinson in the Cambridge debate) which is demotically known as niche marketing (the small business that does not compete but holds to its comparative advantage as a seller of shoes or a bakery, a convenience store or an off licence. This too has a meimetic implementation in the system of ideas.

We begin our deduction of the market value pie, the total value in an economy. This can be seen as a central idea in economics, from Ricardo to Marx, value/ rent/ surplus / normal profit/ supernormal profit are a treated by us as a singular concept.

Growth Accounting, where a production function of labour, capital and value added are fitted to data show that over 70% of growth economically is from technological development. Our dialectic of this is to consider advertising, brand capital, idea map of company and country, as another indicator of value.

The synthesis of growth accounting (supply) and advertising (demand) is in the concept of value, if you like the total amount of money made.

The meimetic structure of capitalism creates rents (value held from competition and competing down of). Capitalism seeks rents as its central mechanism so the market keeps to the structure of five forces model and niche marketing.

On the other hand contestation reduces value, as competition and vying for greater sales leads to lower value per firm. This contestation comes in many less developed countries in the form of miemetic business choices, so for example everyone is in the textile industry, rather than each company keeping its space between each one in the sphere of value, production and advertising.

A model of the less developed country's value total and the meimetic structure of capitalism

$$GDP = Y1 + Y2$$

GDP is gross domestic product (how much money an economy makes) while Y1 is the income from oligopolies and Y2 is the income from niche marketing business'.

As you can see in the meimetic structure of capitalism, GDP is the total value, and is higher than in the less developed country structure:

$N1$ = number of firms who are oligopolies
$N2$ = number of firms who are niche marketing

Then $GDP = N1 \ Y1 + N2 \ Y2$

The LDC case is given as an interrelation between Y1 and Y2, where niche marketing firms (the non oligopolies) become a function, $Y2 = f(Y1)$.

The value total of the two countries are :

<u>GDP</u>
N1Y1.   = value per firm in rich country   (1)

<u>GDP.   .</u> = value per firm as more niche firms start to compete in oligopolies markets (2)
N1Y1+N2Y2

Clearly (2) is lower than (1) as more firms compete and create the contestation to drive down value in the total economy.

<u>The Laplacian and interdependent equations</u>

∇^2 = partial derivatives for each axis of the function. This does not account for interdependency, between the different partial derivatives. You might say that existing mathematics falls short of full solutions and solutions to more kinds of equations.

<u>The Hamiltonian from a canonical perspective</u>

The Hamiltonian canonical equations are:

$$\frac{\delta p}{\delta t} = \frac{\delta H}{\delta q}$$

$$\frac{\delta q}{\delta t} = -\frac{\delta H}{\delta p}$$

This can be seen as a symmetry and balance equation (between shyla and charibus from the Greeks). One is the opposite of the other. A ship travelling between two constraints (islands in the Odessey of Homer) will make changes in course of smaller and smaller difference, a kind of dampening of the oscillating equation of the ship's trajectory. This pont de capiton has been applied to many of physics' theories in our previous book, "Symmetry and Balance, a mathematics book, by Tahir Iqbal". The core structure of Hamiltonians (from a canonical perspective) is an increasing function and a decreasing function that are put together and make an equilibrium, an answer of optimality from mathematics applied as an assertion that does not give credit to the structure and structures of reality.

Essay question: Where does a Hamiltonian exist?

<u>GLS problem of interdependent errors is solved by multiple eigenvalues</u>

Consider the GLS problem (generalised least squared regression), where there are errors, ε_1, ε_2 and ε_3. The sum of ε_i (all errors) is taken to be zero at the solution to the GLS equations.

There are problems with solving this. As ε_1 goes higher, ε_2 should go lower, (as it is in an ordinary least squares regression (OLS) for establishing the maximum likelihood of the coefficients and parameters of the OLS resgression.). Yet ε_3 could share the change with ε_2 so going higher or lower to rebalance the errors to be summing to zero.

There are two ways we suggest, one is of estimating eigenvalues of the errors in their interrelation. Then there is the problem of dividing down the answers to the GLS error estimation and relations.

This can be done by considering a vector field to be solved by genetic algorithms (GA).

The initial estimates and empirically found relations of errors, give an initial distribution of the solutions to the GLS equations. The higher magnitudes in the vector field are selected, whilst lower magnitudes are left out of the solution. This leads to a splitting of the potential candidate solutions. Then a scattering (according to some assumed probability distribution) is made of each solution. In other words where there is an answer , the ball around it is added to the solution set. Then the process goes back to the start process of fitness function (keeping higher magnitudes), splitting and scattering again and again. An assertion needs to be made for when to stop this process as it really needs a way of ruling out mythical solutions (ones that could mathematically exist but not necessarily in terms of our existing knowledge). This added on and found would allow even very complex and large data sets could be dealt with. We as yet have no general answer to the idea of category of myth.

A survey of political economy

Adam Smith in his work, the Wealth of Nations (p.351, 1776) talks about what we see as the pont de capiton for political economy as value added/ surplus/ profit/ rents/ rate of change of money flow. He makes the point of quality and quantity, much echoing Aristotle, (where the idea is species and genus). Quality can be seen as part of value added, as higher quality has higher per unit price, while quantity is the market size. Thus price and quantity in later equations that multiplied together (put as a measure) are the total amount of value. Value added can be seen as the costs of production less the price times quantity. Here then, Smith considers production to be productivity (the physical size of value added) , where a process of capital replacing labour raises labour productivity. At the same time there is the need for investment of quality control of production.

We can see the production process as being labour and capital put together, and additionally, the change in labour and change in capital as being productivity. The change in price is a function of sales (price times quantity).

 Forecasting sales is then a precursor to modelling the economy. This is a challenge to economists who would detract from a heterodox view of economics, can you forecast the economy's sales from all firms, can you forecast an industry sector, can you forecast even a single firm you run yourself?

The obvious way to forecast sales is to take sales in one time period and sales in the previous time period to have an expected difference of zero. The change in sales (hopefully growth) is taken to be a martingale (where the expectation of future is the same as previous periods).

$$P_t \, Q_t \, . = P_{t-1} Q_{t-1}$$

Then there is the problem of there being no new investment and also no depreciation.

An answer to this is dynamic optimisation, the maximising or minimising of an integral. The integral here is an analysis of the sales, $P_t\,Q_t$.

$$P_t\,Q_t = \int dp\,dq \;\rightarrow\; \frac{dPQ}{dp} \;\text{ and }\; \frac{dPQ}{dq}$$

(Here the integral is differentiated, a la dynamic optimisation.)

Doing a Hamiltonian on this leads to :

$$\frac{\delta p}{\delta t} \;=\; \frac{\delta H}{\delta q}$$

$$\frac{\delta q}{\delta t} \;=\; -\frac{\delta H}{\delta p}$$

Then considering this as a measure, of price (p) times quantity (q), you can draw out a series of rectangular or square blocks increasing with the gradient of the ray in these measures determining and setting the relations of supply and demand, which is the Hamiltonian of the sales forecasting equation.

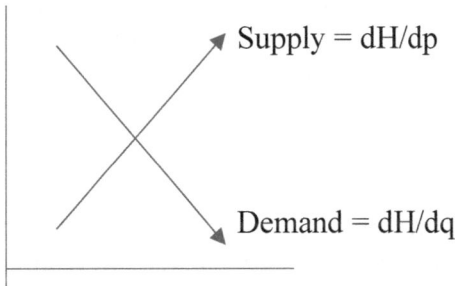

For the price series measure , $\frac{dPQ}{dp} = \int \quad dp$, the figure 1.0 shows the changes to this integral for dp (the change in p), with the ray from the origin showing the gradient of the supply function above. The series of rectangles for quantity (q) are the opposite of the price measure, in the negative of both axes from the price series measure.

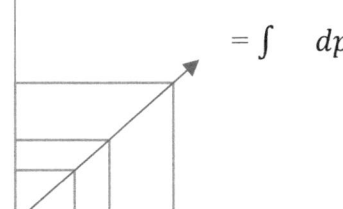

Figure 1.0

Value added from Adam Smith

Smith drew the distinction between productive and unproductive labour. We summarise this as value adding activity and value destroying activity. Smith took it that a household servant does not make value added whilst a worker in a factory produces value added.

We see the concept of value added as the centre of capitalism. Value destroying is bankruptcy and foreclosure of homes, asset market collapse, crime, taxes and overcompetition. However, value destruction is necessary for the price mechanism to work to clear all markets (from overcompeitition) and reseting the economy (asset market collapse).

On sparse matrices and the first and second law of complex systems theories

We take the model for a protocapitalist country of many separated people who do not normally trade with one another. This is known in the mathematical literature as a sparse matrix.

Each node in the economy trades with each node, the transference of money from one person to another depicted as a kind of jacobian matrix, where x1 is the first node, who only trades with x10, whilst the others do not trade with anyone. This is a world of scarcity and Adam smith (WON, p.13) said that such a scarcity leads to a rise in prices and then supply. Thus a sparse matrix becomes more connected, as more people interact in their trades and productions and consumption with each other. Centre nodes develop, and some even find that their own money makes more money. Capitalism is a tremendous mobiliser of resources. People turn to rent seeking, chasing value in the economy and it becomes what could be called developed, rich or a superpower.

Consider Edgeworth's box, the Friday and Robinson Crusoe economy of two people, where one has more the other has less, for n agents. How do you connect the 4 corner edgeworth's box with all other agents? This is not satisfactory so leads to the justification of the model of the economy as jacobian transference matrices. You can summarise this as a global model of a dynamic system, given by for example, 3 nodes trading with each other, yet feedback occurs, so the model of money flowing is an iterative map:

$X(t) = X(t-1) + \theta X(t-1)$

This solves when applied to an infinite limit to be $x^{\theta t}$, either converging on infinity or minus infinity. This is for interdependent equations and is a general result, so called 'first law of complex systems theory'. Given the argument against this of sparse matrices, where there is little or no interdependency, we include the 'second law of complex system's theory', where the onus on the policy maker or capitalist is to plug people into the system of money. Not just equality of opportunity, but successful interventions to give people the social, symbolic and human capital to succeed. Further, to bring people in through mentoring and developing a vibrant and diverse economy and society. Freedom's chains need the release and escape from poverty and the roads to it.

Another answer to the sparse matrix problem for the first law of complex system's theory, you can answer the problem of n agents edgeworth box by making chains of these boxes, each empty corner put with another set of two agents, etc. Clearly you can't be sure to have an even number of agents in society.

However, you can also make small jacobian transference matrices of firms and people who work for those firms, with sub matrices of the trade of a worker (she works for a company as a director but then spends money on a coffee which someone else has money from), done in chains. Thus the sparse matrices are divided up to model the flow. Note you can put any equations and functions you like in this kind of jacobian matrix chain and its repesective nodes.

How to make lots of money

An interesting part of political economy is the theories and models of how processes can be found and used to make increases in economy. For example, there is the process of banking capital, where the increase in assets of an individual allows them more ability to borrow, whilst the increase in assets leads to more money for banks to lend out. Clearly one can see the feedback in this, where there have been so many debt fuelled booms in the economy. The problem of when the chicken has to repay the farmer's generosity of food and lodging, is the point for policy maker on balance. Notably this banking capital process shows that variation in interest rates lead to changes in economy, much avowed by monetarist economics as something that only makes inflation, rather than effect on economy. Note the monetarist interpretation of the transience of interest rate policy which is asserted by them to only affect prices (inflation) is the opposite of our first law of complex systems theory, where the economy grows and inflation is not considered, given the huge productive capacity of the world.

Another process that goes back many centuries is the existence of wood around the world which can be made into ships, then trade occurs which leads to government growth through tariffs, that then connects the world into the first law of complex system's theory.

Consider one for poorer nations, where people make sea salt from the sea at coastal areas, then bring this amazing natural fertiliser to agricultural peoples inland, making roads and money for government to tax and make roads, money and ports with the network of traders (the sea salt makers and distributors) making the basis for a newly developed economy.

The division of time, book II

Introduction

$$x_t \longrightarrow x_{t+1} \longrightarrow x_{t+2}$$

$$y_t \longrightarrow y_{t+2}$$

$$z_{t+1} \longleftarrow z_{t+2}$$

Figure 1.0, different intervals of an equation system.

Figure 1.0 shows our concept of varying intervals, much like jumping of cellular automata between different cells. X causes the next time period and then the next one, whilst y causes the function after two time periods. What z is in figure 1.0 is the expectation of z at time t+1.

This concept is the core of our argument with regard to economics where we say that all economists, especially in neo-liberal and mathematical economics, are not rigorous enough in their mathematics. There is a challenge to the profession to put the structure and axiom of figure 1.0, the varying interval, in respect of the marginalist revolution, where economics looks at elasticities (a common division or group in mathematics). With varying intervals, there are many elasticities so neoliberal economics is not unique and perhaps, not optimal.

As a general example in number theory, we look at prime numbers up to 47.

2 3 5 7 11 13 17 19 23 29 31 37 41 43 47

Figure 1.1 a list of prime numbers

Looking at an interval of 1, the prime number minus the next prime number in figure 1.1 you obtain a series of numbers which are not ameanable to analysis. However, change the interval of the minusing, of the difference between the prime number in the series and the second number within the series you obtain in the majority of cases the difference 6 or 10, with one outlier at $37 - 29$ and $5 - 2$. Thus there is more information in the case of an interval of prime numbers of 2 rather than 1.

We call this interval analysis.

The metric as a place economics is not rigorous enough in

We argue that neoliberal economics has not been rigorous enough in its mathematics. The argument is that there is a mixed up notion of the metric. The metric is the geodesic in a subset of the equations being looked at. It is given as a measured point from a function.

The triangle inequality is an example of a metric, in addition you can heuristically look at it as pythagorous's theorem, that the geodesic (the hypoteneuse) is when squared equal to the other two sides of a right angle triangle squared and summed.

For a simple economic system, you have the well known Marshall's shears, supply and demand curves. The metric (the subsets of it) of supply curves is each unit of the product (its value in production), whilst, for a demand curve, the unit is the use value, the utility or happiness you obtain from a product.

Thus supply and demand are as units different so the intersection of this is a different metric and thus is non-convex. In this argument we see that marginalism, elasticities and differentiation cannot be used because there is no metric; supply of units of product and a model of production is different to use value of a product (its demand).

With optimisation equations and processes used throughout economics there is then no way to solve them, as we have shown the non-convexity of the inability to find marginal quantities (where you have a limit or differentiation or even integral for dynamic optimisation). This is not to say that the mathematics is wrong, rather the mathematics has to be adapted to the exact structures of the problem being looked at.

Consider the problem in dynamic optimisation of PQ, the price times the quantity which determine a market or perhaps industry, product or even macroeconomic variables (total GDP in a country).

PQ is a square integral, on a graph of price (P) and quantity (Q). This can be put into a lagrangian as a constraint which is a subset of the PQ square. This subset is smaller square in the integral. The PQ square can be seen to go to infinity, as an optimisation of PQ. Alternatively, the converging answer (the one most often reported in the literature) is where the lagrangian constraint of cost is equal to the PQ integral, thus meaning that the answer is that there is no profit, PQ = lagrangian constraint.

Consider also P1Q1 as PQ for product 1 and P2Q2 as the integral for product 2. These are each squares (areas on a graph), yet putting them together falls into the problem of different metrics.

For example :

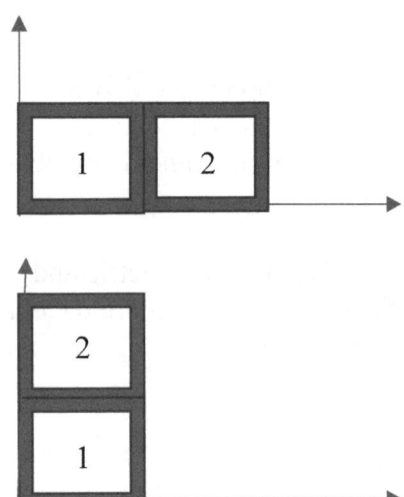

Figure 1.2 Difference in metric for two products, product 1 and product 2 as dynamic optimisations

Seeing figure 1.2 you can see that there are many ways to add the two products together (as P1Q1 and P2Q2), thus a metric (geodesic) has many answers, and therefore the possibility of maximum utility is compromised.

In stochastic analysis of economic equation systems, we note that randomness is simply a scattering (a series of dots randomly on a graph). You can have many metrics for this, as one point on a scattering is related to another, yet for others it is related differently. We discuss scatterings later and develop this idea. You can look at the random variable having arbitrary intervals, especially in time.

Consider the optimisation function for Cass-Koopman's model of utility for a household:

$$\int \frac{C^{1-\sigma}}{1-\sigma} e^{-pt} \, dt$$

σ is the intertemporal marginal of substitution. The metric here is of time and consumption, as units of analysis, which are different, one subset of time is a second or minute or hour or day. One unit of consumption is £10 or £20, for example. Adding in a lagrangian constraint, the λ of the lagrangian are the marginal link to consumption. This is argued to also be of differing metrics. Thus dynamic optimisation of a household utility function is not possible.

Precursors to the capitalist state

An example of intervals in economics and political economy is the structure of the prior to capitalist development and the take off and growth of economies. There is t+1 which is when the market has developed interdependency and has regular growth and there is the interval prior to this, t. We show a number of examples of this which cannot be put together as marginalism, the convexity of the function and model of the world.

1. Fuijan/Hokkien/Taiwan have a history of carpentry, which we believe leads to the introduction of the factory and industrialisation.
2. Chiniot in Pakistan also has a history of carpentry, which is where the feudal/capitalist class come from of Pakistan.
3. Malaysian political economy leads to the wealth of Chinese business being plugged into the malays (the ruling majority in terms of power rather than money). So there are grants and help for farmers who are malay, including the requirement for Chinese business to have malay partners. We can see malaysia's take off in development as being a function of the dispersal of wealth to make a larger market size.
4. East Asian success is through there being large networks of Chinese traders, with online trading giving the connection of this to the world. Thus these linages from traders make value added.
5. Argentina historically was a very rich country, one of the top 10 in the world, yet fell away and collapsed for much of the modern times, possibly because of large amounts of silver mines being extracted and then silver coming into money supply and then inflation and collapse or perhaps debt being built up in the economy which makes larger fluctuations (a high time and a low time but over a long period- see the interval idea).
6. North Korea industrialised before south korea, which was mostly agricultural. Yet north korea is a communist state. Then south korea developed in opposition to this from international political economy.
7. Gershenkron's argument of the need for value adding investment rather than value destroying investment (e.g. from financial market collapses). The point of investment, say a country buying capital every month or one large sum every year is solvable as an equation with interval analysis. The Asian financial crisis caused perhaps by international capital used interdependency in financial instruments and derivatives to make a feedback that lead to the economies collapsing. Study of people from financial markets showed me the concept of reflexivity, which I later formulated into the concept and tools for solution of interdependent equations. Intervals can actually be solved by interdependent equations, which I go into later.
8. The myth of corruption structures the reality of relative sign that is the less developed country (LDC). Everything is a lie because the word library can be rendered as lie-lie (by taking the first syllable of the word 'library' and repeating it).
9. Japanese development was from military defeat from the more technologically advanced Dutch, which lead to MITI, the Japanese government organsiation that facilitated Japanese technological catch up.

Clearly there are many arguments and debates one can have about the above points of what shifts the LDC to becoming rich, thus we see that the concept of intervals changes each argument, each prescription for development economically and politically.

Marshall – the firm as a internally competing products

We put forward the point to economics to develop a theory of forecasting sales. Not only is this something real and practical, it also intervenes in many economic debates. Can an economist's model predict accurately the future sales of a company? We show an answer to this later on in the book, on advertising theories which we develop there.

Firstly, our assumption is that you do not look at the category of the industry, but rather the category of the product. So you can look at a chemical industry or you can look at different products; cleaning sprays, baby food, chocolate.

Economics regularly misses out on this structure, putting everything together as the market, and then finding they are too general answers.

Marshall said (principles of economics, p.315, 1920) that surplus (value added) is from advantages of the firm in equilibrium. We can see this advantage as productivity (though then consider the role in value of advertising).

Productivity of a product can be rising, falling or staying roughly constant. Another one is oscillating productivity. These are all the main analyses of mathematics.

However, we can look at ratios of product's productivities over time. This is a dynamic, where 1 product falls while the other rises. Higher productivity leads to higher surplus and then higher reinvestment to Marshall.

Now consider a dynamic optimisation of a simple function from Marshall:

Profit = Capital x productivity – costs(labour)

Equation 1.0: marshall's equation for profit

Since higher profit leads to more reinvestment and thus capital, profit rises for the most productive firms and most productive products.

A dynamic optimisation of equation 1.0:

$$\int capital . productivity\ dt - costs(labour)$$

But with interval analysis:

$$\int_0^1 reinvestment + \int_1^2 reinvestment + \int_2^3 reinvestment < \int_0^4 reinvestment$$

Where reinvestment is the ratio of profit to more capital reinvested into the product. This is higher for the single investment at time 4 because the regular investment for times 0->1 and 1-> 2 and 2-> 3 compounds and makes higher profit.

This also goes into the concept of the short run and long run. We see economists have already seen intervals in that but do not consider all the many intervals, linkages that can occur and be used in analysis.

Rational Expectations

Consider the concept of rational expectations. This is where the expectation of a variable, say how much the price of a share is on the stock market is expected to be in a year's time, when markets clear, where all information about the asset is priced into the variable. So for example, there has been good news about a stock, then it goes up in value. People often find this as it is often impossible to beat the market at growth in a portfolio.

The expectation of the price is therefore its average, which are equal in rational expectation models (Multh, Econometrica, vol 29, no 3, pp315-335, Jul 1961).

This is obvious from scattering models as a basis for probability theories. See figure 2.0 below:

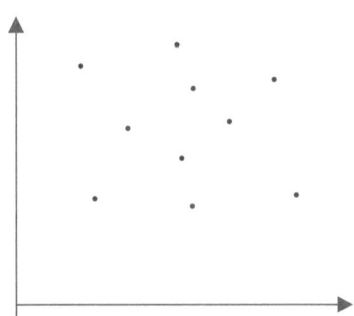

Figure 2.0 a scattering graph for probability theory

Each point of data is randomly shown. As a scattering (that is what we call this generator of ideas), there is perhaps little we could see in its analysis apart from its average (its expectation) and variance, size of dispersion of the scattering data.

However, say you take a piece of data and another one from the scattering, and work out the average between that. You will have a different number to when you take different points in figure 2.0 as differences and averages. Thus probability is unleashed to have many answers.

Thus expectations are not unique.

In addition, consider Muth's theory as divided into different prices at varying periods (varying averages from the scattering). Your expectation at one time or another is a matrix of very many linkages between price time period and expectation (1, 2 or more periods of when the expectation exists).

Given people are trying to beat the market in expectations, this leads to a complicated problem in modelling the market (each person has different expectations due to their own model of the market and the model of the market in the market itself).

We give an answer to this interval problem in expectations later on in work on miemetic expectations.

Kalecki (Journal of Nonlinear mathematical physcis, 8 sup 1, pp 266-271, "The Kaldor-Kalecki model of business cycles as a two dimensional dynamical system"

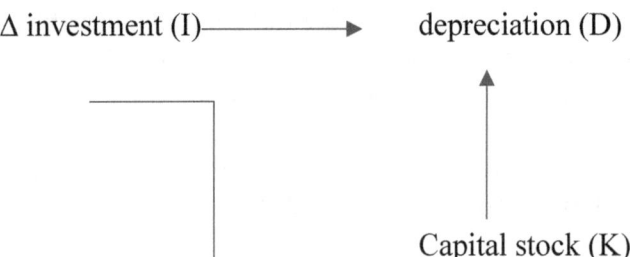

Figure 3.0, oscillator generator of investment

Then :

$$I = K - D$$

Then as the money is reinvested:

$$I(t) = I(t-1) + K - D$$

I(t) is investment at time t, I(t-1) is the interval of the investment.

Then

$$I(t) = K(t) - D(t) + K(t-1) - D(t-1)$$

Looking at what's important in the equation system of investment, capital and depreciation;

	D= f(K)	(1)
And		
	I = g(K-D)	(2)
And		
	ΔK = ΔI	(3)

These previous 3 equations or linkages are amenable to analysis from Jacobians for interdependent equations.

	I	D	K	
I		(2)	(3)	
D			(1)	= Jacobian interdependent equation analysis
K				

This can be solved with Lie group's weyl weights analysis to give long term asymptotes and short run dynamics. You can model one run of the equation system on data by (2) x (3) x (1), crown and thread method, which is made equal to A.

Then look at a time filtration, as A (a kind of interval already used in much of mathematics, e.g. stochastic theories) of the system;

$$\left(\int_0^A D = f(K) + \int_0^A I = g(K-D) + \int_0^A \Delta K = \Delta I + \int_0^A K \right) = A$$

This is an example of how to solve the problem of intervals, yet it is a nebulous subject and very interesting as an argument against neo-liberal economics. However Kalecki is also amenable to the analysis.

Note there is the concept of market clearing in the static phase of analysis in economics, where the market has no waste, and is equal to PQ (price times quantity). Yet for economic growth to exist above the previous level, there must be a positive different in PQ of the whole economy:

$$P(t)Q(t) > P(t-1) Q(t-1)$$

Therefore you can't have the price/market mechanism of allocation of resources, demand and investment returns because of economic growth. This is a way of looking at Kalecki's above model.

Ramsey model and economic growth
(Ramsey model, p73, Phase space figure 2.1, Economic Growth, Barro and Sala-i-Martin, 1999)

The marginal utility elasticity in Ramsey's model is $-\theta > 0$, of utility against consumption. The interval problem is easily seen as the basis for an argument of axioms of intervals, as follows.

A woman buys a coffee every day. This is consumption in terms of product as objective function. They also buy a dress every 7 days. In addition she owns a house and pays interest payments on the mortgage which she pays every 3 months. Clearly the model of utility's relation to consumption is dependent on what time period we look for, what interval we are basing our calculations of.

So C(t=i) is consumption at time i.

Then we see the utility function has non-unique values:

C(t=1) = 1 cup of coffee.
C(t=2) = 2 cups of coffee.
C(t=7) = 7 cups of coffee and 1 dress.
C(t=3 months) = coffee, dresses and one repayment of mortgage loan

What's interesting in Ramsey's model is the phase space diagram. We employ and develop a theory later of phase space diagrams, if you like, generators of the dynamics of the nonlinear systems and interdependent equation systems. Where Ramsey finds a saddle point in the phase space (in other words an equilibrium for central parts of the answer) we believe in all is flux, that the foot is never put in the same river twice (Heraclitus, Fragments). This phase space we are investigating because of its amazing property to solve special theories of interdependent equation systems, is from the Kammaa equation (see earlier on in this book)

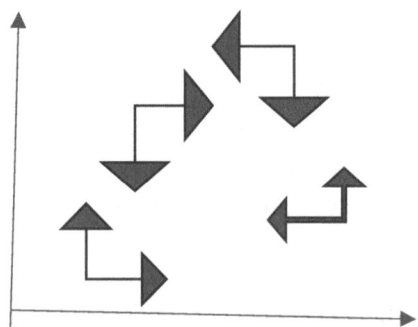

Figure 4.0, Kammaa equation phase space as an alternative to Ramsey's model

We can also look at this as planck's box, from the theory in physics of black body radiation oscillating and bouncing inside a constrained set.

The sum of the differences in this from Kammaa equations is:

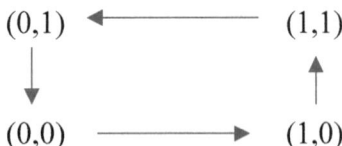

Figure 4.1 Planck's box and kamma equations

Each data point goes around in a normalised circuit. Ramsey's model does not show this, instead solving as exponentials then moving inside that and coming up with their phase-space diagram.

Note the sum of the data points in the phase space of figure 4.1 is equal to (2,2) but as intervals of (1,0) from (0,0) = (-1,0), (1,1) from (1,0) = (0,-1), (0,1) from (1,1) = (1,0) and finally (0,0) from (0,1) = (0,1). These numbers all add up to (0,0). This gives an orbit of the function's trajectory. Interestingly you could use a sine wave to simulate and model the planck's box of oscillations.

The kamma equations could be applied to the ramsey model with planck's box as a major generator of the model. Note it is still in keeping with the concept of equilibrium as shown by the sum of data points.

Dynamic optimisation is the differentiation of an integral. It is often about maximising or minimising an integral. It is highly regarded, for example it is how aeroplane's wings function to create lift.

Can dynamic optimisation deal with the following problem? You sell your goods at the beginning of each time period in a 3 node economy (x1,x2,x3) , x(i) is the value of goods you sell. Each node sells to each other, with feedback happening, giving solution that the ratio of money being transferred between nodes is d time x(i-1).

Then consider layers of trades in time, x1 sells at time t, x2 sells at time t+4 and x3 sells at time t+2.5

Thus we see the point of the division of time as an axiom. How intervals in functions at asserted times that are not the same lead to a different jacobian / transference matrix. This applied to the whole analysis lead to endless trees buzzing through the economy with connections and linkages growing and falling as value is created in this.

How to develop the Ramsey model? Simply by using interdependent equation system tools (e.g. Jacobian) applied to intervals. Intervals are interdependency.

Quality control problem

X = g Q

A = cost of checking an item is faulty.

X is the value of returns that are faulty and returned to a producing company. G is the ratio of returns from production and Q is the total quantity produced.

This is a very interesting problem in interdependent equations, which show why there are troublesome conclusions to it.

At first blush you might solve this problem of finding out how often to check quality of goods in a production line in a factory, such that you maximise profits, through the following equation:

Minimise – Returns x Cost of Checking = X A

With E(X) being known then:

E(X) A = X A

So

A = g

Then Price x quantity – g A = Profit

Sales – Return costs = cost of checking faulty item
➔ PQ – X = A

But X = g Q

If you have E(x)= average, then $E(X)=\frac{1}{n}\sum x$, which further complicates the problem shown here as interdependency (the Q and X are endogenized, are pulling the system in different directions and are involuted into them, thus giving many answers but also an ever circling dynamic of solution and deduction).

Then you can have multiple equilibria, with different high and low values against each other of the different variables.

We take the concept of interdependency and show how you can solve this and intervals as the object of this, in addition to bringing to light a huge new paradigm in everywhere, the interdependent revolution.

Answers to the interval of time

Hypothesis testing with Interdependent equations

Kammaa equations, the spiral and w-shaped asymptote can be modelled as ARMA and GARCH processes. ARMA is autoregressive moving average, while GARCH is generalised autoregressive conditional heteroscedasticity.

A kamma equation looks like this:

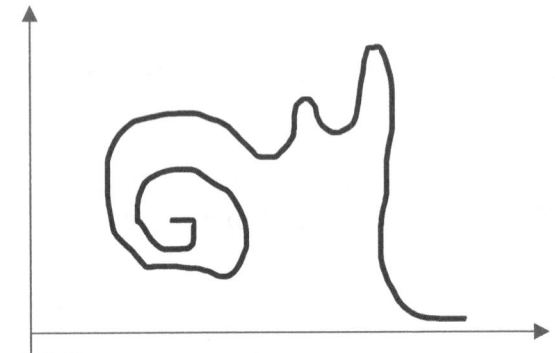

Figure 5.0 Kammaa equation

What can be seen in figure 5.0 is the autoregressiveness of autoregression.

Autoregression is the returning to the average of a statistical function. Done in itself it is possible to see this as two variables, one dying out at the higher parts of kamma and leading the previous spiral (small variance) part of figure 5.0 to become a rise, fall, rise and then large fall. We have tried investing in a share portfolio following this and have good returns, predicting when the share will break out of a small band of variance. The autoregression at the spiral of figure 5.0 is applied to itself, making the small spiral. Many random processes and data can be put into this structure.

The current Kantian (theory of science and proof) is A causes B, one variable causes another. This is the basis for scientific experiments by have a control to examine the randomness in A causes B having a correlation and an experiment for A put with B.

So A -> B   (experiment)
And $\varnothing$ -> B   (control experiment)

P(A) | P (B) = linkage of causality in independent equations, which is estimated in the experiment.

However, many systems are not simply one variable causes another. With interdependency a number of variables (B, C, D) cause and generate the data of hypothesis testing.

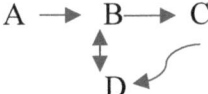

Figure 5.1 interdependent system hypothesis testing.

In figure 5.1 A cause B, which causes C and then D, yet D causes B and B causes D. As a corollary of the independent equation thesis, you are looking for a conditional probability of:

P(A | B , C , D)

This can be seen in set theory as the intersection of set A with B, C, D. This is then needed to be estimated, so

Y(t) = A x(t) + B x(t-1) + ε(t) + ε(t-1)

A and B are different weights for the model of autoregression. The change in Y(t) is modelled so where B x(t-1) > A x(t) then dY(t) is falling, whilst when B x(t-1) < A x(t) then dY(t) is rising. This is a way to write an autoregressive equation.

This is the AR part of ARMA. The MA, moving average part of ARMA statistical analysis is easily shown as:

Y(t) = x(t) + ε(t)

With the maximum likelihood (expectation/ average) of x is :

$$E(x) = \frac{1}{i} \sum x_i$$

We can say applying ARMA to Kammaa equations that:

Moving average part of Kammaa:

B = C + D + A

(for figure 5.1 system, where it is solved at B, B is the crown of this crown and thread analysis).

Autoregressive part of kamma:

B = α C + β D + χ Λ

This autoregressive part of kamma can be involuted or run again and again as cycles of calculation of crown and thread. The crown is the orphan node that leads to thesis, antithesis and synthesis. Consider the flow from A in figure 5.1 going to B through to D, (one cycle) and then back to B, which itself is higher from more A flow. At some point as A runs out of flow, there is a spinning cycle of B to C to D to B again. This is understood as the thesis-antithesis-synthesis as the process spins out on itself to very high values, as seen in kamma. We also see the flipping, as B goes to D as well, and the whole system moves in the opposite flow for a while. This is through a very basic example of interdependency (figure 5.1) so is possibly quite general and accounts for the fluctuations so often used to in Fourier analysis.

In respect of GARCH as a way to find the numbers for Kammaa equations, using data from real systems, consider the relative sign (difference) of two scatterings, one with large variance ($\sigma1$) and one with small variance ($\sigma2$). This gives the concept of conditional variance (but note not covariance), where x (t) -> $\sigma1$ or $\sigma2$.

The autoregressive part of the GARCH model can be seen as x(t)= A x(t-1), where A>1 then x(t) is rising, A=1 then x(t) is unchanged and A<1 then x(t) is falling.

Then GARCH model is:

$$x (t) = f(\sigma) + A\, x(t\text{-}1)$$

then where:

$$A\, x(t\text{-}1) = f(B\, x(t\text{-}1))$$

Is an autoregression of an autoregression (AR), as the AR processes are made functions of the parent AR process.

You can model A's relation to B by:

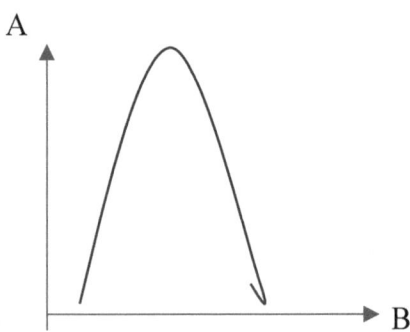

Figure 5.2 AR in AR

As B's effect dissipates, so A's flow takes over and makes the kamma equation go from initial spiral to asymptotes that are autoregressive (go up and down).

Note, non stationary time series analysis must be eschewed, as extracting a trend from the data to make non-stationarity into stationarity takes away information from the analysis since there is an intersection of the set of data and the set of the trend estimate.

In addition you can use genetic algorithms to model and forecast and perhaps control the interdependent equation system. We mention this in our book (The division of time, book 1, Tahir Iqbal 2021) as an answer to vector fields. You can analyse the Kammaa equation and also interdependent equation analytically with vector fields. This is not novel but linking GARCH and ARMA's respectable status to interdependent equations which kamma especially expouses and perhaps generates, gives hope to those looking for a deeper and more general reality.

<u>Hamiltonian with many variables of interdependency</u>

The Hamiltonian is a way of estimating and modelling functions. From the canonical approach there is a collecting together of equations and twisting round in the differentials of those functions. Going back to figure 5.1, we see that we can analyse this as a Hamiltonian using jacobian transference matrices of interdependency.

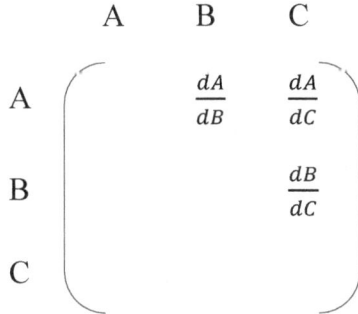

$$
\begin{array}{ccc}
A & B & C \\
\end{array}
$$

$$
\begin{array}{c}
A \\
B \\
C \\
\end{array}
\begin{pmatrix}
& \dfrac{dA}{dB} & \dfrac{dA}{dC} \\
& & \dfrac{dB}{dC} \\
& & \\
\end{pmatrix}
$$

Figure 5.3 Jacobian transference matrix for interdependent equation system where the hypothesis of a complex system is being estimated

We do not look at the bottom triangle of the matrix in figure 5.3 as this is simply the opposite transference for the variables shown in the top right triangle (where the differential ratios are put). In addition it is not necessary to look at A->A, B->B and C-> C as these are if non-zero commutativity and mean a feedback in the system from itself, leading to an exponential. Interestingly enough, where the differential ratios given above in figure 5.3 are above 1 in summation, there is also an asymptote of infinity. Where the summation is less than 0 then the asymptote of a flow in the differentials is minus infinity. If the total summation is 1 then the system converges, has a non-infinite value, yet even in the infinity there is huge scope to introduce the concept of a special solution to equations of interdependent systems. This is the simplicity of the once unexplored territory of complex systems, probability theory, chaos theory, nonlinear dynamics, iterative maps and bifurcations and oscillating structures.

We call the case of solving the asymptotes of figure 5.3, as lie group weights. If the summation is over 1 then the whole matrix can be replaced with the function e^{xt} . For summation less than 0 then the general solution is e^{-xt}.
The e^{xt} is the lie group weight. Many mathematicians throw out any equation that is non-converging (not becoming infinity like e^{xt} does). Yet inside this there is a special solution, given by kamaa equations (of AR in AR processes and dynamics, consider a sine wave in a sine wave as used in fourier analysis).

How to control such a system? This is very important in being useful to the engineer or practitioner of mathematics. The literature on coupled nonlinear dynamic systems is really saying the system is x(t) and the coupling of this is x(t) again, as x(t) – x(t) = 0, therefore control of it.

The reason such ideas do not work is because many systems are much like figure 5.1, where there is interdependency, here put as a simple version, with reality being many layers of this, yet with the lie group weight analysis, there is the potential for an answer.

Giving way to the literature, the control of the figure 5.1 system is an equal and opposite system. Using a multidimensional Hamiltonian gives an idea of how to solve this, as you have:

$$\frac{\delta H}{\delta A} = \frac{\delta A}{\delta B} \delta t$$

$$\frac{\delta H}{\delta A} = \frac{\delta A}{\delta B} \delta t$$

$$\frac{\delta H}{\delta D} = \frac{\delta D}{\delta B} \delta t$$

These solutions to the Hamiltonian can be made into a vector field :

$$X = \frac{\delta H}{\delta A}, \frac{\delta H}{\delta A}, \frac{\delta H}{\delta D}$$

Note there are in fact many ways to write a multidimensional Hamiltonian so we appreciate if the reader finds different results.

Then you find changes in the vector fields' differentials using genetic algorithms. You can see which direction of the equation is overriding and dominating relative to other values in the vector field.

Application of interdependent tools to LDC (Less Developed Countries) debt problems

An LDC (sometimes derogatively called the 'Third World') are often seen to be beset with debt crises that lead to austere and punishing policy especially experienced by the poor. We show how the nature of this problem is interdependency and its solution is also that.

The equation for LDC debt is:

Debt = Balance of payments (B) + Government budget deficit (G) + exchange rate (S)

We assume the following relations in the structure we are analysing:

S= f(debt). , as debt outflows are money outflow from the country which reduces demand for the domestic currency of the LDC.

B=g(S), balance of payments is a function of exchange rate, where differing export and industrial structures lead to higher balance of payments deficits from falling exchange rate from debt outflows in the previous equation.

G= debt + B, as debt repayments go high so government needs to run a budget deficit to repay debt.

Debt outflow -> change in S -> change in B-> change in G -> change in debt

Note we are primarily looking at government debt, though we leave the reader to find the system for all debt held in the country.

You can do a jacobian transference matrix of this, following our elucidation of this previously.

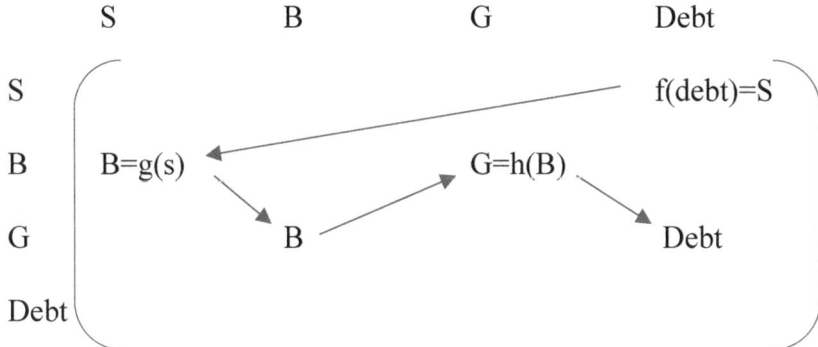

Figure 5.4 system of equations analysed as crown and thread.

What is interesting is how policy can be formed from this to make linkages that are different to existing policy. For example, you could change industrial structure to be value adding, such that when the exchange rate falls from debt outflows, there is higher exports. You could for example, link the government budget deficit to the exchange rate by government buying domestic currency on the exchange rate. Another idea is forex bonds, where people buy bonds on the back of balance of payments level. Also more taxation from tariffs can lower the budget deficit and solve the balance of payments problem of non-value adding exports.

An interesting aside to the analysis is that you can also determine eigenvalues of the matrix of jacobian transference. Doing this, you can see that you can have multiple equilibria, for example, having high debt and high G and B, or low debt and low G and B. Also you can have low debt and high G and B, and high debt and low G and B.

How do you do this with intervals? The answer is to use a technique of differences of intervals.

For example, consider the series: 1,2,3,4,5,6

The differences between these intervals are (e.g. 2-1, 3-2, 4-3)

Series x: 1   2   3   4   5   6

$\delta x =$ +1   +1   +1.   +1.   +1

$\delta x^2=$.   +0   +0   +0   +0

You can consider the differentiation of intervals as being able by this difference method that can make differentials up to any magnitude.

Division of time can also be seen as a factorisation of the equation system into its multiples/ factors. So function $d(mn) = d(m) \, d(n)$.

Applying division of time (intervals as an axiom) to a torus would seem to give the definition of charts (freeze frames in the torus) summed together. Each chart is a way of expressing the interval in this advanced formulation as a torus.

A torus can also be seen as a collection of cones (a doughnut split into ice cream cones). However, with interdependency, the cones can be out of phase with each other, so overlapping and overlapping the overlap. (Torus actions on varieties p89-90, Algebraic Quotients, Torus Actions, Correll, McGuiness (ed.) 2002). Note in this book lie groups are seen without interdependency which is actually the main reason for them.

A mobius strip shows a way of doing the topology of interdependency, so looking at torus' will also show the analysis, (as torus' are multiples of mobius strips).

Forecasting sales

The challenge to economics is to predict sales (turnover or amount of sold production) in a time interval. Analytically, sales are equal to the price multiplied by the quantity, so are clearly there in existence in many areas and analyses of economics.

The key point is that rent seeking (the price mechanism looking for surplus or high profits, which is necessary for markets to clear or be in equilibrium) requires a forecast of sales, in order to make the needed investment decisions and employing of labour.

A rent is untapped value, potential profit, urging supply from firms of it. Rent seeking is thus a central feature of the market.

We consider the analysis of the rent as being divided into each different product a firm makes, with the potential for competition inside the firm, as one product overlaps another in demand.

Looking at Say's law, that supply creates its own demand, we can view the concept and institution of advertising to be the reality of this law. Advertising creates demand / sales of a product, which is already given in its existence. Advertising makes value added/ rents.

You can look at different kinds of ratios and mathematical structures to analyse and advise advertising. For example you could look at the return (profit) of an investment from advertising. You can also look at the return from advertising (the change in sales and from that expected sales) from a firm's capital (its share price and value of the company). You can also look at the macro advertising variable, the average value added from advertising for a series of competitors or industry or even entire economy.

To develop more tools to forecasting sales, you can try to estimate the probability of sales given advertising. You would then need to create experimental indicators, through comparing this probability with the probability of not sales from advertising and probability of sales given no advertising. Making the average for a firm allows for analysis and developing real actionable strategies for the advertising campaign means higher profit and coming closer to the answer of forecasting sales.

Advertising can be seen as creating a rent;

Advertising as rent/ value added:

```
<barrier to entry.
<reputation among peers>
<information and use value>                    ⎤
<quality and cost, exchange value>            ⎬      <A>
<efficiency through creating economies of scale>
<becoming the dominant member of the market>
<increasing value added and rent>             ⎦
```

Each of these things can be added on by a firm's research department and also by the economist to develop a database of value added as advertising and making the model for the marketing department of a firm.

From <A> you could obtain the indicator of:

$$P(S|A) = f(ROI) = Pi(S|A) - P(S|A) \qquad (1)$$

(1) Is the relative return compared to the industry or peer level of return from advertising.

The above is a mathematical analysis of advertising. We look at some sociological and semiotic theories in this section. Gramsci holds the idea of the interest group making the semiotics that lead to advertising's success or failure. There is the interest in the environment, so the structure of language and conversations is such that there is value in electric cars or solar panels. There are in fact many interest groups in many countries, so there is some cause to be aware of heterogeneity when making research in the area of signs/ semiotics.

Hegel shows the idea of engulfing, so that power brings into itself any opposition to it. Predator firms take over and push out of the market, smaller companies. Advertising as power and bringing in counter-cultural elements also makes the social consciousness as language and signs. The colour green is a sign of environmentalism and is thereby something of value. Marxist thought of ideology is also clear to see as power changing semiotics. Alienation, the Marxist idea of separation from capitalism, where people feel sad leads to the privileging of comedy and laughter signs. This is also salved with empowerment, as you choose more of what you want, rather than a prescriptive monologue.

From Lacan, you can see the point of semiotics giving meaning, necessarily in the sign's definition. Advertising as psychotherapy and giving meaning, the meaning of life, is another way to forecast sales (change in the probability of sales to advertising). Foucault and De Beauvior can be looked into to see the semiotic prison of reality, the structure of the society, as being broken out of in art, thus the acceptance of a product from what makes the sign system (which the advert associates itself with). The gestalt from signs, whereby you feel happy, say for example, equality so the person who is the relative sign is given a privilege as advertising as justice. What is also important is that we are often receivers of the system of signs, rather than transmitters of the sign and message.

All these theoretical approaches can be leveraged to make consensus for buying the product whereby the advertising is seen as a big success.

Where is interdependency in this?

The firm can be seen as production and sales with the use of a product and the idea map of a product. Marketing departments can be put into a correspondence of this as Quality of the product, popularity (sales) linked to information about a product and its place in a system of relative signs.

Putting this rubric with interdependent equation system's tools (jacobian transference) you can easily apply this to develop interdependency and the infinite state automata to make expanding of sales at a lower cost of advertising.

Lacan (Ecrit, On question prior to possible treatment of psychosis) showed this insightful model of the metaphor in semiotics:]

$$\frac{S}{S\prime} \cdot \frac{S\prime}{x} \longrightarrow S\left(\frac{1}{s\prime}\right)$$

Figure 5.5 Lacan's equation of metaphor

In figure 5.5 we see the concept of the non-existent sign, S' as an interpretation (metaphor) of metaphor. The equation can be seen as :

Captialism (S) is craziness (S'); its symptom (1/s) and cure (S) are consumption (S) and profit (1/s).

This statement is beautiful because it uses advanced tools of semiotics, where there is an involution in the art. Consider then the conversation tree, that someone replies to the statement with:

Advertising is a balm for capitalism in making being out of the relative sign (1/s).

And then:

Narcissism is too much ego. (as a dialectic). Looking into the mirror as a feminine in a clothing shop, the catwalk, the marriage.

From this we can see the concept of the conversation tree, the development and play of signs in a system not of the central controller of the novel but rather the real and meaningful course of everyday life in terms of conversation.

Now consider the interdependency of signs. You can have condensed signs, where there are multiple meanings for multiple people, and with conversation trees an impact on what people say in dialogue and among many. Poetry is an example of condensed signs, where Keats talks of the beauty of the nightingale's song and mortality and finishing of the beauty of life.

Another concept of particular note for forecasting sales (and creating more on the balance sheet) is the concept of the non-existent sign being transposed and displaced by a sign, the unknown becoming unknown. This is especially interesting in educational theory as the non-existent sign is the origin of children learning, in analysis, as the development of a sign to play into the non-existent.

Consider the classic Saussurean sign as signifier and signified. S/s. You could displace S with a cup of coffee and s as a table. The coffee is on top of the table. This gives utility as it is easy and relaxing to drink the coffee. However, in interdependent analysis of the conversation tree (the dialectic of signs from Saussure) you could have two people with the same link, having a cup of coffee with each other. Thus we see how interdependency is informing and useful, even in semiotic analysis.

The composition of the non-existent sign, a poem

The non-existent sign is the orphan
It has no history as a sign
It may be that it is a sign closed down (revolutionary)
Or it may be something too popular (over hyped)
It is something no one has thought of (the name that cannot be named)
S -> S/S' -> S'

It is the dialectic of S (not S),
      In a teleological world (the future is not the past)

- Linguistic concept of a metaphor as a metaphor of systems

Importance: what is important is the non-existent sign
Nothing is the non-existent sign
         → Everything is the sign
: so what's important is nothing and everything

There should be a general attraction to the non-existent sign
:everything is attracted to nothing and everything.

But what makes difference? And what relations does this engender?

The reader is invited to complete and compete with the following analysis of the society towards the sign system:

MONEY, POWER, BEAUTY, SOCIAL CAPITAL, MYTH AS ORDERING, FUN, MORALITY, FAME,SEX, IDENTITY, LEISURE

Each one of these is an eigenvalue/ point of attraction. Your marketing model of advertising can be built out of research on different products in history and their salving, their use as a balm to the sign system above.

Foucauldian analysis of the conversation

Hegel	The ideal	Dialectic
Hitler	The message	Miemetics
Saussure	The sign	S/s, the sign
Levi-Strauss	Structure	Language as myth
Lacan	Post structuralism	The metaphor
Tahir Iqbal	interdependency	Conversation trees

Each phase of history shown above is an attrition of the non-existent sign. Yet the painting that has a picture of a pipe with the writing 'ceci n'est pas une pipe' shows an involution that destroys the sign as a non-existent sign. Is the non-existent sign myth? And therefore the absence of myth. Spin as the myth of government, the absence of you is the non-existent sign of government!

The miemetic network of consumption

You can view an economy as a system of nodes (each person) where there are linkages between each one in respect of consumption choices. So for example one person wears a miniskirt, then they are seen by another who then goes on to wear a miniskirt. Given certain properties the whole system may become miemetically wearing the same clothing style. Then the system flips and everyone is wearing jeans. Another product, the fur coat is being worn, but eggs are thrown at them for harming animals, so the fur coat does not flourish.

You can use fourier analysis, perhaps simplifying this to looking at fluctuations and modelling of them with cosine waves, such that cos(t) is the amount of sales you have, as cos(t) divided by the number of people, you have the average probability of the sales.

You may find that the market becomes saturated for a product most of the time, the rising star becoming cash cow and then an old product that is not having its initial lustre. This is modelled as a dampening out of the sine wave model of a miemetic network of consumption (of sales). This is amenable to a spiral in kamma equations, which forestalls big fluctuations at the end of the product's dominance in the market.

If you obtain data on sales and their time series, you may see that you can analyse this with an attractor, from nonlinear dynamics equations. Once you have found the attractor for the product you are looking at you can then produce actionable research by looking at what changes the attractor. For example, an eigenvalue of low value in terms of sales could be optimised to show how to obtain infinite sales.

Consumption models as nonlinear dynamics

$$C = y + \dot{y} + \ddot{y} \quad (1)$$

This shows that consumption (C) is a function of y, dy/dt and d2y/dt2; the variable, its ratio with respect to time and its ratio in respect of respect of time.

You may put consumption , C, as cos(wt), from our previous point of fourier analysis of sales above.

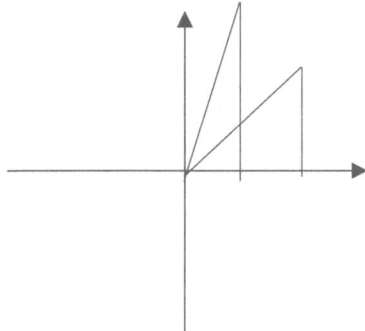

Figure 6.0 the ratio of right angled triangles that create a cosine wave (ratio of adjacent length to hypotenuse)

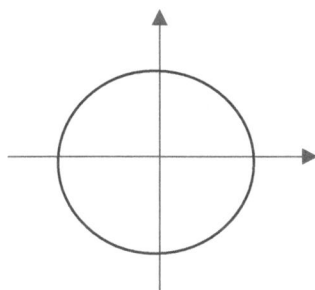

Figure 6.1 the limit of the triangles that are right angled forming a circle around 2π radians

Figure 6.0 shows the change in right angled triangles which can form a circle at the end point of the hypotenuse shown in figure 6.1. This is why there are fluctuations, as the cosine's angle is higher, then so is the ratio of them, going round in an equal circle in figure 6.1.

Because of this you can argue that consumption in this model is cos(wt) and so when substituted into equation (1) becomes the well known duffing oscillator in nonlinear dynamics. This is a very interesting equation to develop an understanding of as it gives a maturity to one's mathematics.

Amplitude of the consumption equation (1) is given as a variable A multiplied to the cosine wave cos(wt). We can model a dampening of the consumption variable with for example:

Cos(1) = 1 cos (wt)
Cos(2) = ½ cos(wt)
Cos(3) = ¼ cos(wt)

This is an example of modelling kamma equations, where a central spiral is modelled as the dampening of cos(x).

The duffing oscillator shown above can be used instead of the finite horizon consumption model of economics, which does not give a clear idea of the dynamics of sales = consumption. Typically in finite horizon analysis, one maximises a utility function subject to a savings function. The problems found in quality control previously apply as well, as consumption is causing saving which is causing consumption, thus being non-commutative and thereby amenable to interdependent analysis.

Miemetic Expectations from Keynes

(Keynes, The General theory of employment, interest and money, p.317, 1936)

Previously unanalysable concepts like Keynes's idea of animal spirits of the economy, where firms are going together as a herd to be pessimistic about the state of business or are too optimistic, as booms happen then, and then recession, are possible to have more light shed on them with interdependency. Keynes states that the marginal efficiency of capital falls off in the oscillation of the sum total economy. So the change in capital invested is conditional in its probability on the amount of return of capital, which falls.

The change in capital invested is caused by market psychology to Keynes. This is a miemetic structure. Doing a fourier transform on the total money and assets in an economy, given as PQ (price times quantity subdivided in the total money and assets), you can have the result:

$$\int P(t)Q(t)e^{2\pi t} \, dt$$

This can be transformed into sine and cosine functions to model an out of phase oscillating function (for example $e^{2\pi t} = \sin(t) + \cos(t)$, where the sine wave and cosine wave overlap, in other words are out of phase).

Out of phase functions, for example out of phase cones in a torus or overlapping parabolas are a cutting edge mathematical tool to create real answers to mathematical problems.

The out of phase can be seen as a delay or echo in the function's trajectory. Intervals here therefore overlap, given the intersection of sets in the model.

A concluding remark on Assets for the poor

Household model reinvented:

Wages -> household capital -> assets for the poor

A household which has higher wages from labour monopoly power (given helping the poor is a moral high ground) makes higher surplus from Adam Smith's wealth of nations, p.164, Book I-III).

As they have higher wages they then have higher household capital from surplus from monopoly power of labour. The price component of the commodity or good go higher from higher wages. Thus GDP which equals sales (PQ) is rising. This then feeds back into household capital which then is used to buy assets for the poor household, raising the value of assets, the rent disassociated with labours needs to maintain surplus). Thus there is a feedback interdependency.

That is our answer and also can be seen as history in modern times of economic growth in assets and then fall.

The author would like to thank Pink Pavilion guest house for sponsoring this work and providing support for its research and publication

Tahir Iqbal

61